Geometry Fundamentals: Exploring Shapes and Spatial Relationships

GEOMETRY FUNDAMENTALS: EXPLORING SHAPES AND SPATIAL RELATIONSHIPS

Prof. Avanish Kumar

ARCLER PRESS

www.arclerpress.com

Geometry Fundamentals: Exploring Shapes and Spatial Relationships

Prof. Avanish Kumar

Arcler Press
224 Shoreacres Road
Burlington, ON L7L 2H2
Canada
www.arclerpress.com
Email: orders@arclereducation.com

ISBN: 978-1-77469-939-3 (Hardcover)

ABOUT THE AUTHOR

Prof. Avanish Kumar is presently working as Professor at the Department of Mathematical Sciences and Computer Applications, Bundelkhand University, Jhansi. He has completed his M.Sc., from Meerut University (presently CCS University), Meerut in 1989, M. Phil., from University of Roorkee (presently IIT), Roorkee in 1992 and Ph. D. in Mathematics, from Gurukula Kangri, University, Hardwar in 1999. University of Roorkee awarded him 'University Medal' for standing First in First Class. He has served as Assistant Scientific Officer in the Commission for Scientific and Technical Terminology, Ministry of Human Resource Development, Government of India, New Delhi w.e.f. 20 November 1995 to 25 November 1999. He joined Bundelkhand University as Assistant Professor in the Department of Mathematical Sciences and Computer Applications on 26 November 1999. There he served in various capacity and became Professor on 26 November 2014. He served as Chairman of the Commission for Scientific and Technical Terminology, Department of Higher Education, Ministry of Education w.e.f. 06 February 2017 to 05 February 2021. He was having additional charge of the Director, Central Hindi Directorate, Department of Higher Education, Ministry of Education, Government of India w. e. f. 07 July 2017 to April 08, 2020. He has supervised 14 Ph.D. thesis, 28 M. Phil. Dissertation. He has published about 90 research papers in National and International Journal of reputes. He is the author of two books and editor of four terminological related books. His research interests are Modeling, Optimization Techniques and Distributed Computing Systems. He is the life member of International Academy of Physical Sciences, Indian Mathematical Society, Computer Society of India, Operations Research Society of India, and many other similar academic bodies and societies. Also, he is the Vice President, of Operations Research Society of India (ORSI) and Vijñāna Parishad of India (VPI). During 2010 to 2013 he was the Head of the Department of Mathematical Sciences and Computer Applications. During the same period 2010 to 2013 he was also the Head of the Institute of Computer Science and Information Technology. He was the Chief of University Employment Information and Counseling Centre (UP Government) Bundelkhand University, Jhansi during 11 June 2009 to 05 February 2017. During this period Ministry of Employment and Labor, Government of India, New Delhi awarded the establishment of Modal Career Centre (MCC) under the University Employment Information and

Counseling Centre (UP Government) Bundelkhand University, Jhansi to facilitate the job seekers of this university as well as Bundelkhand Region with career related trainings, guidance and counseling. He was visited to the Rajamangla University of Technology. Thanyaburi (Thailand), University of Science and Culture, Tehran (Iran), Kathmandu, (Nepal), Port Louise (Mauritius), Shanghai (China), Guadalajara (Mexico) for various academic work. In 2000, he was awarded 'Professor R. C. Mehrotra Best Teacher Award and Gold Medal' for outstanding contribution to the university services by the Bundelkhand University, Jhansi and in 2016, Bharat Vikas Award was conferred to him by the Institute of Self Reliance, Bhubaneswar, Odisha, in appreciation of dedicated services to the society and humanity.

TABLE OF CONTENTS

LIST OF FIGURES

LIST OF ABBREVIATIONS

2D	Two Dimensions
3D	Three Dimensions
AA	Angle-Angle
AR	Augmented Reality
ASA	Angle-Side-Angle
AU	Astronomical Units
CAD	Computer-Aided Design
CT	Computed Tomography
DEM	Digital Elevation Model
GD&T	Geometric Dimensioning and Tolerancing
GIS	Geographic Information Systems
GPS	Global Positioning System
HL	Hypotenuse-Leg
LSA	Lateral Surface Area
MRI	Magnetic Resonance Imaging
NURBS	Non-Uniform Rational B-Spline
SAS	Side-Angle-Side
SI	International System of Units
SSS	Side-Side-Side
VR	Virtual Reality

PREFACE

Geometry, at its essence, explores the fundamental principles governing shapes and their relationships in space. Its origins date back to ancient times when early scholars started studying and understanding angles and lines. Discipline of geometry has undergone a remarkable evolution over thousands of years, progressing from the rudimentary geometric studies of ancient civilizations to the structured and rigorous system of knowledge famously formulated by Euclid. Today, geometry occupies a central position in mathematics and extends its reach into various practical domains. Its time-honored wisdom serves as a guiding light for professionals in architecture, enabling the design and construction of buildings that stand as marvels of structural ingenuity. In engineering, geometry is the bedrock upon which intricate designs for bridges, machines, and electronic circuits are built.

The following pages present "Geometry Fundamentals," a comprehensive examination of geometry's foundational aspects.

Chapter 1: "Introduction to Geometry" commences by exploring the historical development of geometry and its cultural significance. It discusses basic geometric shapes, geometric constructions, measurements, units, and the relevance of coordinate geometry in plotting points and lines on a plane.

Chapter 2: "Fundamentals of Shapes" expands on geometric shapes, classifying them into two-dimensional, three-dimensional, and irregular categories. It also delves into the properties of shapes, including dimensions, angles, and symmetry. Concepts like congruence and similarity are explored.

Chapter 3: "Spatial Relationships and Transformations" discusses spatial aspects of geometry, covering distance, length, collinearity, and coplanarity. The chapter also examines spatial visualization skills and their applications in computer graphics and geographic information systems.

Chapter 4: "Coordinate Geometry" introduces the Cartesian coordinate system, its application in plotting points and lines, and equations for lines and curves. The chapter highlights the relevance of coordinate geometry in physics, engineering, and computer science.

Chapter 5: "Introduction to Similarity" focuses on understanding similar figures, exploring similarity transformations, proportional figures, and methods for proving similarity. The golden ratio and its significance in art and nature are also discussed.

Chapter 6: "Three-Dimensional Geometry" extends geometric principles into three-dimensional space. It covers coordinate systems, distance, equations for lines and planes in three dimensions, vectors, and transformation matrices. Applications of three-dimensional geometry in various fields are presented.

Chapter 7: "Geometrical Models" discusses geometric abstraction in modeling and introduces wireframe, surface, and solid models. The chapter also explores the role of computers in geometric modeling and provides practical problems for further comprehension.

Chapter 8: "Applications of Geometry in the Real World" brings together the knowledge acquired in the preceding chapters and highlights geometry's relevance in fields such as architecture, engineering, medicine, computer graphics, and environmental sustainability.

Throughout this book, readers will find diagrams, examples, and exercises aimed at facilitating the comprehension and application of geometric concepts. Whether you are a student seeking to build a strong foundation in geometry or a professional looking to apply geometric principles in your work, "Geometry Fundamentals" offers valuable insights and knowledge.

CHAPTER 1

Introduction to Geometry

Contents

1.1. INTRODUCTION

Geometry, derived from the Greek words "geo" (Earth) and "metron" (measurement), is the mathematical study of space and the properties of objects within that space (Trofimov, 2013). It encompasses many topics, making it a multifaceted discipline with practical and theoretical applications. Geometry can be broadly categorized into two main branches:

1. **Euclidean Geometry:** Named after the ancient Greek mathematician Euclid, forms the foundation of classical geometry. It deals with the properties of flat or two-dimensional shapes, such as points, lines, angles, triangles, polygons, and circles (Brodmann & Sharp, 2012). Euclidean geometry is governed by a set of axioms and postulates introduced by Euclid in his work "Elements."

   Euclidean geometry extends to concepts like parallel lines, congruence, similarity, and the Pythagorean theorem. It provides the basis for understanding the properties of two-dimensional space, making it indispensable in fields such as architecture, engineering, and art (Marques et al., 2003). For example, consider the Pythagorean theorem, which states that in a right-angled triangle, the square of the hypotenuse's length is equal to the sum of the squares of the other two sides. This theorem has numerous real-world applications, from calculating distances to designing stable structures (Mathai, 1999). Mathematically:

   $$a^2 + b^2 = c^2$$

2. **Non-Euclidean Geometry:** It explores spaces with properties that differ from those in Euclidean geometry. There are primarily two types of non-Euclidean geometries (Eisenhart, 2015):

   i. **Spherical Geometry:** Where lines are great circles on the surface of a sphere.

   ii. **Hyperbolic Geometry:** Characterized by negatively curved surfaces.

These non-Euclidean geometries have applications in astronomy, relativity theory, and map projections. Spherical geometry, for instance, finds relevance in navigation and the study of celestial bodies, while hyperbolic geometry challenges our understanding of curved spaces. The development of non-Euclidean geometry marked a significant departure from classical geometry, leading to new insights into the nature of space (Minsky & Papert,

1969). For example, the Haversine formula calculates the shortest distance between two points on the surface of a sphere, crucial for navigation and astronomy. In the intricate dance of angles, trigonometric functions, and the Earth's radius, this formula:

$$hav(q) = sin^2 \frac{q}{2}$$

This equation encapsulates the essence of spherical geometry. Join us in this chapter as we unravel the mathematical intricacies of spherical surfaces, exploring their unique properties and unleashing the power of equations to navigate the intricacies of our three-dimensional world.

1.1.1. Historical Development of Geometry

The history of geometry spans thousands of years and involves contributions from various cultures. Geometry's roots can be traced back to ancient civilizations such as the Egyptians and Babylonians, who used geometric principles for practical purposes like land surveying and construction (Figure 1.1). However, the ancient Greeks elevated geometry to a formal discipline. Euclid's "Elements," written around 300 BCE, is a landmark work that systematically organized the knowledge of geometry at the time. It presented a rigorous axiomatic foundation for geometry and introduced the famous Euclidean postulates, which remain influential in mathematics today (Pottmann et al., 1999).

Figure 1.1. Illustration of Babylonian calculations using triangles.

Source: Michael Marshall, Creative Commons License.

During the Islamic Golden Age, scholars like Al-Khwarizmi and Omar Khayyam furthered the development of geometry and introduced algebraic methods into the field. The fusion of algebra and geometry led to advancements in conic sections (Pottmann et al., 1999). One notable contribution from the Golden Age scholars in geometry and algebra is the quadratic equation, which has roots in both fields. One such quadratic equation can be represented as follows:

$$ax^2 \aleph bx \quad c \quad 0$$

This general quadratic equation, where (a),(b), and (c) are constants, was extensively explored by scholars like Al-Khwarizmi and Omar Khayyam. In the 17th century, René Descartes introduced coordinate geometry, also known as Cartesian geometry, which established a link between algebra and geometry. This innovation paved the way for analytical geometry, a critical precursor to modern geometry (Sullivan, 2001). The 19th and 20th centuries witnessed groundbreaking developments, including the discovery of non-Euclidean geometries by Carl Friedrich Gauss, János Bolyai, and Nikolai Lobachevsky. Albert Einstein's theory of general relativity, which describes the curvature of spacetime, also expanded the geometric landscape (Pomeroy et al., 2019).

1.1.2. Cultural Significance of Geometry

Geometry has not only played a central role in mathematics but has also left a profound cultural impact. Here are some ways in which geometry has influenced cultures throughout history (Pomeroy et al., 2019):

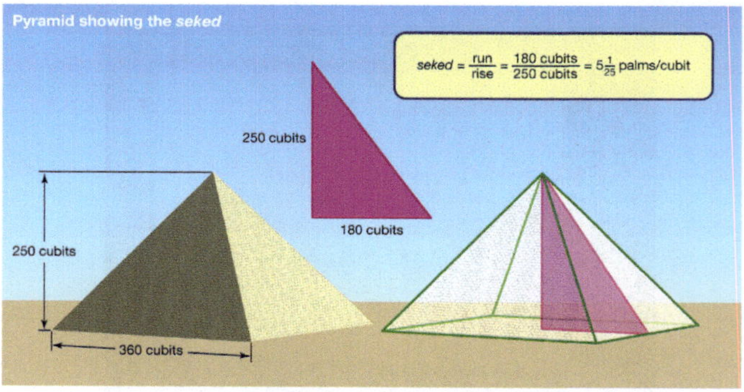

Figure 1.2. Geometry involved in the great pyramid of Giza.

Source: Brittanica, Creative Commons License.

1. **Architecture:** Many architectural wonders, such as the pyramids of Egypt, the Parthenon in Greece, and the Taj Mahal in India,

showcase the application of geometric principles. Architects and builders have relied on geometry to create aesthetically pleasing and structurally sound structures. Consider the construction of the Great Pyramid of Giza, which exhibits remarkable geometric precision. Its sides align almost perfectly with the cardinal points of the compass, and the pyramid's dimensions reflect mathematical constants like pi (π) and the golden ratio (Seidenberg, 1961) (Figure 1.2).

2. **Artistic Expression:** Artists have drawn inspiration from geometry to create visually stunning works. The use of geometric shapes, symmetry, and perspective can be seen in various art forms, from Islamic geometric patterns to the intricate designs of MC Escher. In Islamic art, geometric patterns known as "girih" adorn mosques and palaces. These patterns are constructed using a compass and a straightedge, demonstrating the profound connection between geometry and aesthetics (Wagner et al., 2013). One such example is the use of "Muqarnas," a decorative element where geometric shapes create stunning three-dimensional vaults.

   The mathematical equation for the height (h) of a Muqarnas dome's steps can be expressed as:

$$h = \frac{r}{2} tan\left(\frac{ü}{n}\right)$$

where; (r) is the radius of the Muqarnas dome; and (n) represents the number of steps.

3. **Navigation and Exploration:** Geometry has been essential for navigation and exploration. Ancient mariners used geometric concepts to navigate vast oceans, while astronomers employed geometry to understand celestial bodies' positions and movements. During the Age of Exploration, sailors relied on celestial navigation techniques that measured angles between the horizon and celestial objects. These measurements, based on geometric principles, enabled accurate determination of their locations.

4. **Cultural Symbolism:** Geometry has held symbolic significance in many cultures. For example, the circle, representing perfection and eternity, appears in various cultural and religious contexts.

The use of geometric patterns in Islamic art and architecture symbolizes the infinite nature of God's creation. The mandala, a geometric design often in the form of a circle, holds spiritual importance in Hinduism and Buddhism. It represents the universe, unity, and harmony, showcasing how geometry transcends mathematical boundaries to become a symbol of cultural and spiritual significance (Gangnard et al., 2019) (Figure 1.3).

Figure 1.3. Illustration of mandala floral style.

Source: Vecteezy, Creative Commons License.

In summary, geometry is a multifaceted mathematical discipline with deep historical roots and profound cultural significance. Its study extends beyond mere shapes and figures, encompassing many concepts and applications that continue to shape our understanding of the world.

1.2. BASIC GEOMETRIC SHAPES

Studying shapes and spatial relationships begins with fundamental elements—points, lines, and planes. These concepts form the building blocks of geometric structures and are crucial in solving complex problems in two and three dimensions. This section will study fundamental geometric shapes and their characteristics (Turrin et al., 2011).

1.2.1. Point and Its Characteristics

A "point" in geometry is an elementary object with no dimension. It is often represented as a dot and a fundamental space location. Points are unique in that they are defined solely by their coordinates. For instance, in a two-dimensional Cartesian coordinate system, a point can be described as (x, y), where 'x' and 'y' represent their respective positions along the x-axis and y-axis. However, it is essential to understand that a point has no length, width, or height; it is infinitesimally small (Figure 1.4).

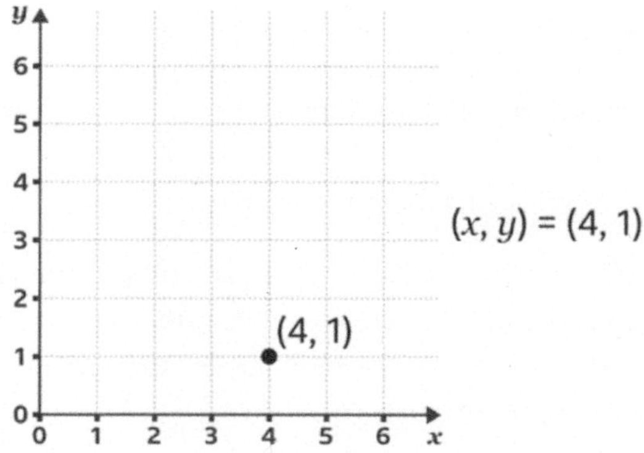

Figure 1.4. Illustration of a point where x =4 and y = 1.

Source: BBC, Creative Commons License.

1.2.2. Line and Its Characteristics

A "line" in geometry is a straight path that extends infinitely in both directions. It can be imagined as a collection of infinite points that follow a particular direction. Lines are characterized by their slope, which indicates the line's steepness, and their equation in various forms, such as slope-intercept form:

$$y = mx + b$$

There are different types of lines, such as horizontal lines with zero slope and vertical lines with undefined slopes. Additionally, lines can intersect, forming angles and shapes (Catterall et al., 1999).

1.2.2.1. Types of Lines

There are various types of geometric lines, each with its defining characteristics:

1. **Straight Lines:** These are lines that do not deviate from their direction. They maintain a constant slope and are the shortest distance between two points. Straight lines are often denoted by linear equations like y = mx + b, where 'm' represents the slope, and 'b' represents the y-intercept (Figure 1.5).

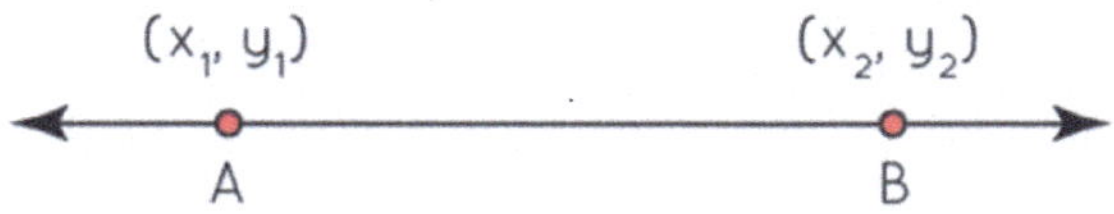

Figure 1.5. Illustration of a straight line.

Source: CueMath, Creative Commons License.

2. **Curved Lines:** Unlike straight lines, curved lines change direction as they progress. A constant slope does not define these lines and can form various shapes, such as circles, ellipses, and parabolas (Figure 1.6).

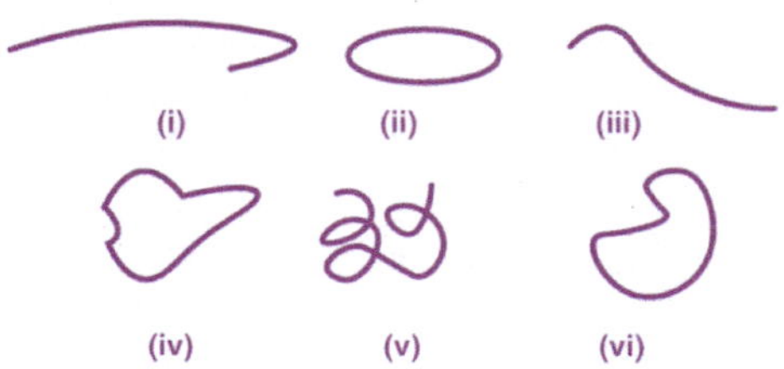

Figure 1.6. Illustration of various curved lines.

Source: BYJU's, Creative Commons License.

3. **Infinite Lines:** An infinite line extends indefinitely in both directions. It has no endpoints and can be represented as a collection of points that follow a specific direction. Infinite lines are crucial for understanding spatial relationships in geometry and can be precisely defined using linear equations.
4. **Line Segments:** A line segment, on the other hand, is a finite portion of a line with two distinct endpoints. Unlike infinite lines, line segments have a measurable length and can measure distances between points in space.

1.2.2.2. Ray and Its Properties

A ray is a geometric object that consists of a single endpoint, called the initial point, and extends infinitely in one direction. Rays are often described as "half-lines" because they have a starting point but continue indefinitely in one direction. Rays can be named using their initial point and any other point on the ray, such as "Ray AB," where A is the initial point and B is any other ray (Bruce et al., 2017) (Figure 1.7).

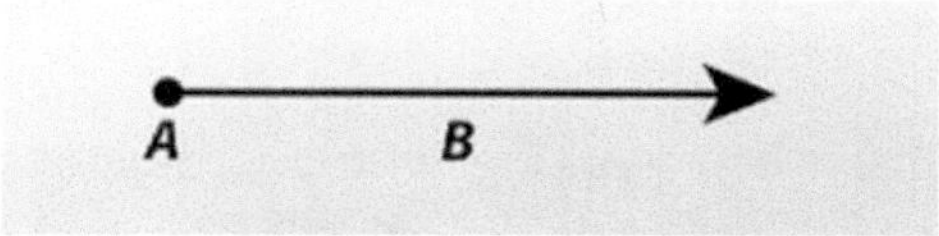

Figure 1.7. Illustration of a ray AB.

Source: Joseph Vigil, Creative Commons License.

Rays, like lines, can also be classified based on their properties. For instance, opposite rays are two that share the same initial point but extend in opposite directions, creating a straight line (Stieff, 2011). Understanding the distinctions between different types of lines, line segments, and rays is essential in geometry. It allows us to precisely describe the nature of objects, angles, and intersections in space. Moreover, these concepts are fundamental when working with coordinate geometry, as they form the basis for graphing equations and solving geometric problems involving lines and shapes (Shu et al., 2023).

1.2.3. Planes in Three-Dimensional Space

In three-dimensional geometry, "planes" play a significant role. A plane is a flat, two-dimensional surface that extends infinitely in all directions. Think of it as a tabletop or a sheet of paper. At least three non-collinear points uniquely define a plane, or an equation can represent it:

$$A_x + B_y + C_z + D = 0$$

where; A, B, C, and D, are constants. Planes can intersect with one another, forming lines, and they can also intersect with lines, producing points (Klein et al., 2020) (Figure 1.8).

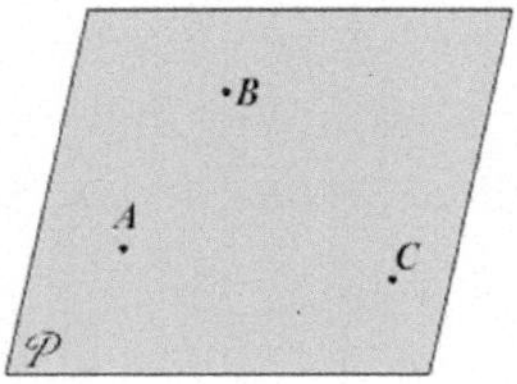

Figure 1.8. Schematic of an ABC plane.

Source: Varsity Tutors, Creative Commons License.

Understanding these foundational concepts—points, lines, and planes—is essential for comprehending more complex geometric structures and relationships. They provide the framework upon which the entire field of geometry is built.

In practical terms, when working with coordinates, lines, and planes, these concepts allow us to precisely describe the location of objects in space, create accurate drawings and diagrams, and solve problems that involve spatial arrangements. Geometry, as a mathematical discipline, relies on the clarity and precision of these fundamental elements to explore the richness of our spatial world (Jacobson & Kitchin, 1997).

1.2.4. Angles and Their Classification

In geometry, angles are fundamental geometric figures formed by two rays or line segments that share a common endpoint, the vertex. Understanding angles and their classification is essential for describing shapes, spatial relationships, and geometric properties (Russ et al., 2008).

Angle is represented by a symbol ∠. For example, an angle of 45° will be represented as follows:

$\angle A = 45°$

In this equation, ∠A represents the angle, and 45° represents the measure of the angle in degrees.

An angle is formed when two rays or line segments share a common endpoint, called the vertex. The two rays that compose an angle are referred to as its arms. The arms of an angle can extend indefinitely, creating different angle sizes and measurements.

1.2.4.1. Types of Angles

Angles can be classified into several types based on their measurements and characteristics (Jafri et al., 2017):

1. **Acute Angle:** It is an angle that measures less than 90°. Its sharpness and openness characterize it, making it smaller than a right angle.
2. **Obtuse Angle:** It measures more than 90° but less than 180°. It appears wider and less sharp than a right angle.
3. **Right Angle:** It is exactly 90° in measurement. It forms a perfect L-shape and is often denoted by a square symbol (∟).

4. **Straight Angle:** It measures 180°, forming a straight line. It is often denoted by a straight-line symbol (—) (Figure 1.9).

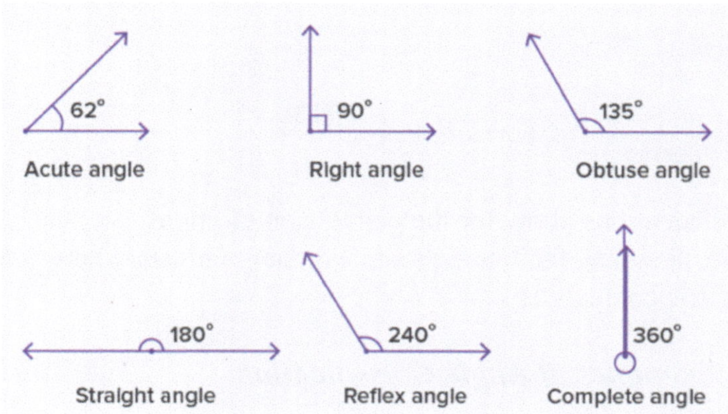

Figure 1.9. Illustration of various types of angles.

Source: Splashlearn, Creative Commons License.

1.2.4.2. Angle Measurement

Angles can be measured in two primary units (Salzman et al., 1999):

1. **Degrees:** The most common unit of angle measurement is degrees (°). A full circle is 360°; angles are measured by the number of degrees they span.
2. **Radians:** These are an alternative unit of angle measurement frequently used in trigonometry and calculus. In a full circle, there are 2π radians (approximately 6.28318 radians). Radians are often used in mathematical equations involving angles, especially in trigonometric functions.

The conversion between degrees and radians is a fundamental concept in trigonometry. The relationship between degrees and radians can be expressed through the following equations:

i. **From Degrees to Radians:**

$$Radian = \left(\frac{p}{180}\right) \times Degree$$

or simply,

$$q\left(in\,radians\right) = q\left(in\,degrees\right) \times \left(\frac{p}{180}\right)$$

ii. From Radians to Degrees:

$$Degree = \left(\frac{180}{p}\right) \times Radian$$

or,

$$q\left(in\, degrees\right) = q\left(in\, radians\right) \times \left(\frac{180}{p}\right)$$

These equations allow for the conversion of angles between the degree and radian measures, facilitating seamless communication between different trigonometric contexts.

1.2.4.3. Examples of Angle Classification

Let's consider some examples to illustrate these angle types:

1. **Acute Angle:** An angle of 45°, like the one formed by a clock's hour and minute hands at 3:00.
2. **Obtuse Angle:** An angle of 135°, formed by opening a book to the midpoint between fully closed and fully open.
3. **Right Angle:** The angle formed by folding a rectangular piece of paper along its diagonal creates a right angle.
4. **Straight Angle:** When two straight-line segments are perfectly aligned, they form a straight angle.

Understanding angle classification is crucial for solving geometric problems, working with trigonometric functions, and describing the relationships between lines and shapes. Angles play a fundamental role in geometry and are essential for comprehending the spatial arrangements of objects in our world.

1.2.5. Properties of Triangles

Triangles are fundamental geometric shapes characterized by three sides and three angles. Understanding their properties is crucial in geometry and is the building block for more complex geometric structures. Triangles have various applications in various fields, from architecture to engineering.

1.2.5.1. Types of Triangles

Triangles can be classified into several types based on the lengths of their sides and the measures of their angles:

1. **Equilateral Triangle:** In an equilateral triangle, all three sides are of equal length, and all three angles are congruent, measuring 60° each. It is the most symmetrical type of triangle.
2. **Isosceles Triangle:** It has two sides of equal length, making its two base angles congruent. The third angle, called the vertex angle, can vary in size.

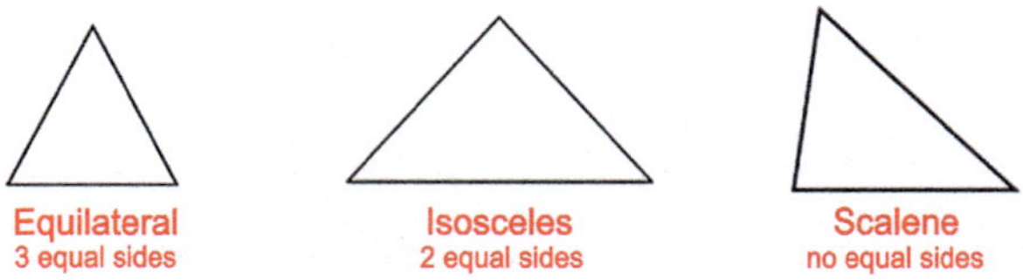

Figure 1.10. Schematic of the types of triangles.

Source: Voovers, Creative Commons License.

3. **Scalene Triangle:** It has no sides of equal length, and all three angles differ in measure. It lacks the symmetry of equilateral and isosceles triangles (Figure 1.10).

1.2.5.2. Triangle Inequalities and Sum Theorem

Understanding the relationships between the sides and angles of triangles is essential. The triangle inequality theorem states that in any triangle:

- The sum of the lengths of any two sides must be greater than the length of the third. This condition ensures that the triangle can be closed and is not degenerate.
- The difference between the lengths of any two sides must be less than the length of the third. It prevents one side from extending too far, maintaining the triangle's integrity.

The triangle sum theorem, on the other hand, relates to the angles of a triangle:

"The sum of the interior angles of a triangle is always 180°."

Mathematically,

$$\angle A+\angle B+\angle C=180^\circ$$

In this equation, $\angle A, \angle B,$ and $\angle C$ represent the interior angles of the triangle. The equation emphasizes that the total measure of the interior angles $\angle A, \angle B,$ and $\angle C$ in any triangle always adds up to 180°.

Its fundamental property allows for the classification of triangles and the calculation of unknown angles.

Understanding the properties of triangles lays the foundation for solving complex geometric problems and is a fundamental skill in various STEM (science, technology, engineering, and mathematics) disciplines. Triangles are the basis for trigonometry, which has countless practical applications in physics, engineering, and computer graphics.

1.3. GEOMETRIC CONSTRUCTIONS

Geometric constructions are a fundamental aspect of geometry, dating back to ancient civilizations like the Greeks, who developed a system of straightedge and compass constructions to create various geometric shapes and solve geometric problems. These constructions use only two simple tools: a straightedge (an unmarked ruler) and a compass. Through the precise application of these tools, geometric figures are constructed with mathematical rigor.

1.3.1. Straightedge and Compass Constructions

Following is the explanation of straightedge and compass:

1. **Straightedge:** It is used to draw straight lines between two points. The straightedge must have no markings, ensuring that the constructions are based solely on geometric principles, not measurements.
2. **Compass:** It is an instrument consisting of two arms, one with a pointed end and the other with a pencil or ink nib. It is used to draw circles and arcs of various radii.

1.3.2. Constructing Basic Geometric Shapes

Following is the construction of various geometric shapes (Ponce et al., 2017):

Tip

Geometric constructions are not just about drawing shapes; they have practical applications too. Architects and engineers use these constructions to design buildings and structures with precision. Understanding their applications can open doors to creative problem-solving

1.3.2.1. Constructing a Line Segment

Following are the steps to construct a line segment:

i. Place points A and B on the paper to construct a line segment AB.

ii. Open the compass to a suitable width and place the compass point on A.

iii. Without changing the compass width, draw an arc intersecting the line. Label the intersection point as C.

iv. Place the compass point on C and draw a similar arc intersecting the first arc. Label its intersection point as D.

v. Draw the straight-line CD. You have now constructed the line segment AB (Figure 1.11).

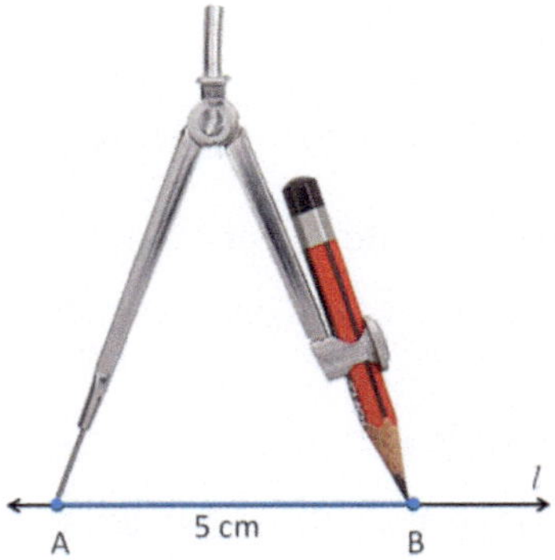

Figure 1.11. Schematic of the construction of a line segment AB.

Source: Teachoo, Creative Commons License.

1.3.2.2. Constructing an Equilateral Triangle

Following are the steps to construct an equilateral triangle:

i. Start with a line segment AB.

ii. Place the compass point on A and set the width to AB.

iii. Draw an arc that intersects AB at C.

iv. Place the compass point on C and draw a similar arc that intersects the first arc at D.

v. Place the compass point on D and draw a final arc intersecting the previous arcs at E.

vi. Draw the straight lines AC, BC, and CE to form an equilateral triangle (Figure 1.12).

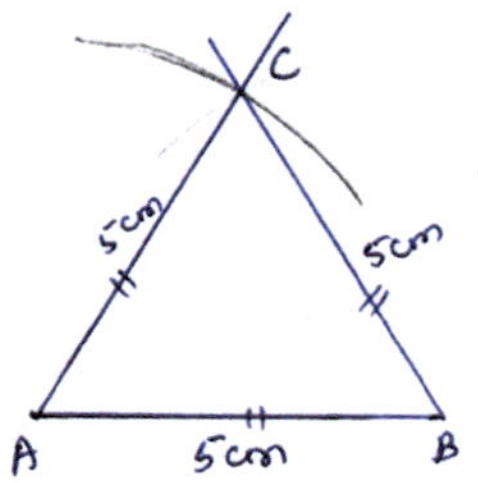

Figure 1.12. Construction of an equilateral triangle ABC.

Source: Toppr, Creative Commons License.

1.3.2.3. Constructing a Perpendicular Bisector

Following are the steps to construct a perpendicular bisector:

i. Given a line segment AB, you can construct its perpendicular bisector.
ii. Place the compass point on A and set the width to over half of AB.
iii. Draw an arc above and below AB intersecting at points C and D.
iv. Draw the straight lines AC and BD.
v. The intersection point O of AC and BD is the midpoint of AB, and line OD is the perpendicular bisector of AB (Figure 1.13).

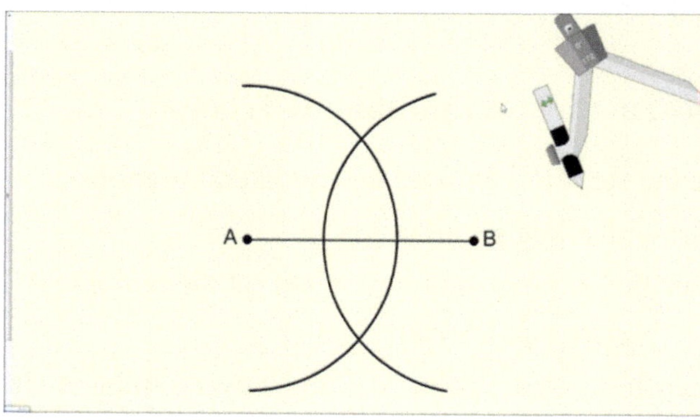

Figure 1.13. Construction of a perpendicular bisector using a compass.

Source: Vinteachesmath, Creative Commons License.

1.3.2.4. Constructing a Parallel Line

Following are the steps to construct a parallel line:

i. To construct a line parallel to a given line L through a point P, not on L.

ii. Place the compass point on P and draw a small arc intersecting L at two points, A and B.

iii. Without changing the compass width, place the compass point on A and draw an arc above L.

iv. Place the compass point on B and draw a similar arc above L.

v. Draw the line through P and the intersection points of the arcs (Figure 1.14).

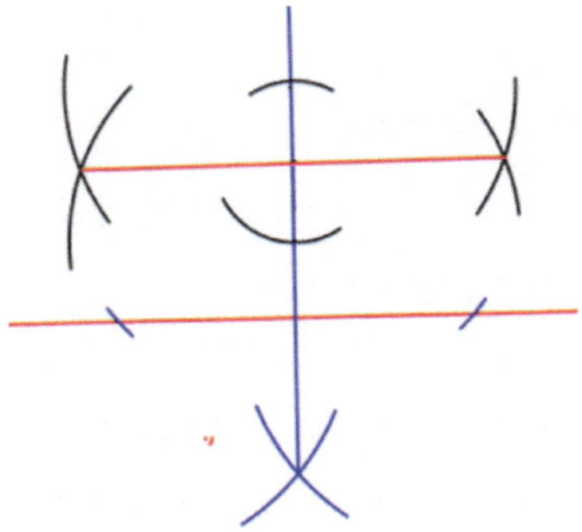

Figure 1.14. Construction of parallel lines.

Source: Instructables, Creative Commons License.

1.3.2.5. Constructing an Angle Bisector

Following are the steps to construct an angle bisector:

i. Given an angle XYZ, you can construct its angle bisector.

ii. Place the compass point on X and draw an arc that intersects the rays XY and XZ at points A and B, respectively.

iii. Place the compass point on A and draw an arc intersecting the first at C.

iv. Draw the straight line XC, which bisects angle XYZ (Figure 1.15).

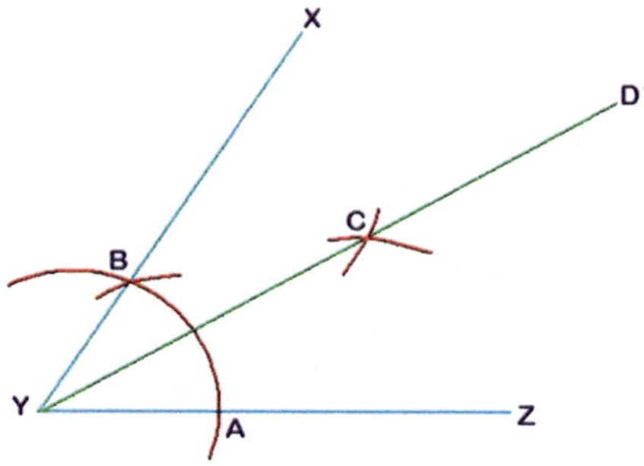

Figure 1.15. Construction of an angle bisector.

Source: Brainly, Creative Commons License.

1.3.2.6. Constructing a Square

Following are the steps to construct a square (Mykytiuk & Shenitzer, 2002):

i. Start with a line segment AB.

ii. Construct perpendicular bisectors for both AB and BA.

iii. Label the intersection of the perpendicular bisectors as O.

iv. Use O as the center to draw four congruent circles, each with a radius equal to OA.

v. The vertices of the square are the points where the circles intersect (Figure 1.16).

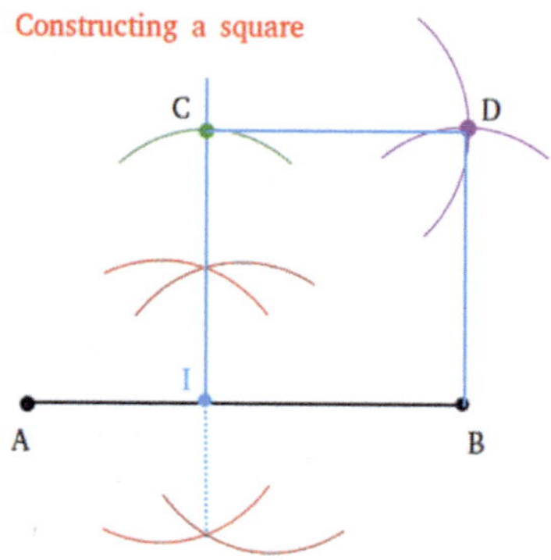

Figure 1.16. Illustration of a square construction.

Source: Basic Mathematics, Creative Commons License.

1.3.3. Applications of Geometric Constructions

Geometric constructions are not only valuable for creating basic shapes but also for solving complex geometric problems. They are extensively used in architectural designs, engineering, and various branches of mathematics, including Euclidean geometry and trigonometry. These constructions provide a solid foundation for understanding and applying geometric principles in practical scenarios, making them an essential skill for students and professionals in STEM fields (Kim & Lee, 2012).

1.4. MEASURING AND UNITS

Measurement plays a crucial role in geometry and the study of spatial relationships. Accurate measurement allows us to quantify the properties of geometric figures, such as lengths, angles, areas, and volumes. To ensure precision and consistency, standardized units of measurement are employed.

These units provide a common language for mathematicians, scientists, engineers, and professionals to communicate effectively and work with geometric data. Whether measuring the sides of polygons, the angles of triangles, or the dimensions of three-dimensional objects, adopting a standardized system of units ensures that geometric measurements are universally understood and applicable across diverse fields of study (Wang et al., 2022).

1.4.1. Measurement in Geometry

Measurement enables us to assign numerical values to geometric properties such as lengths, angles, areas, and volumes, making it an indispensable tool for understanding and analyzing spatial relationships. In this section, we will study the fundamental principles of measurement within geometry, exploring concepts, techniques, and the mathematical language underpinning the field's crucial aspect (Herbst, 2002).

Measuring geometry entails determining a geometric entity's extent, size, or quantity, typically concerning a chosen unit of measurement. This unit serves as a reference point, allowing for the expressiveness of measurements in numerical terms.

For instance, when measuring the length of a line segment, we may employ units like inches, centimeters, or meters, depending on the context and desired level of precision.

1.4.1.1. Length Measurement

Length is one of the most fundamental geometric measurements and extensively applies to various geometric figures. For instance, precise length measurement is essential in measuring the sides of polygons or the dimensions of geometric shapes. The Pythagorean theorem, a foundation of geometry, enables us to calculate the hypotenuse length in a right-angled triangle, given the lengths of the other two sides. This theorem is expressed by (Ho et al., 2016):

$$c^2 = a^2 + b^2$$

where; (c) represents the length of the hypotenuse; and (a) and (b) denote the lengths of the other two sides.

1.4.1.2. Angle Measurement

Angles are another critical geometric property subject to measurement. In geometry, angles are typically measured in degrees or radians. A complete circle is divided into 360°, and each degree is further subdivided into 60 minutes, with each minute containing 60 seconds. This system facilitates precise measurement of angles, allowing us to express angles accurately (Bodduluri & Ravani, 1993).

Mathematically, angle measurement can be expressed as follows:

$$Complete\,Circle = 360^\circ$$

In this equation, a full circle is divided into ü °. Additionally, angles can be measured in degrees, minutes, and seconds. For example:

$$1^\circ = 60'$$

$$1' = 60''$$

These equations signify the subdivisions of angles into degrees, minutes, and seconds, allowing for precise and standardized geometric angle measurement.

1.4.1.3. Area and Volume Measurement

Area and volume measurements are pivotal in geometry, especially when dealing with two- and three-dimensional figures. The formulas for calculating the area of various shapes, such as rectangles, triangles, and circles, are essential for quantifying enclosed regions' extent. For instance, the area of a rectangle is calculated by multiplying its length by its width:

$$A = \text{length} \times \text{width}$$

Volume, conversely, pertains to the measurement of the space enclosed by three-dimensional objects. Formulas for calculating volume are specific to the geometric figure in question. For example, the volume of a cylinder is given by:

$$V = ðr^2h$$

where; (V) is the volume; (ð) is a constant (approximately 3.14159); (r) is the radius of the cylinder's base; and (h) is its height (Cai et al., 2013).

1.4.2. Units of Measurement in Geometry

Units of measurement are a fundamental component of the language of geometry. They provide the framework for assigning numerical values to geometric properties, enabling precise and consistent analysis of spatial relationships. In this section, we will study the units of measurement about geometry, exploring the various systems, their significance, and how they are applied in mathematical and real-world contexts (Rho et al., 2010).

1.4.2.1. The International System of Units (SI)

The International System of Units, abbreviated as SI from the French "Système International d'Unités," is a globally recognized and standardized measurement system. SI units are utilized across various fields, including geometry, to ensure measurement consistency and precision. The SI system encompasses a wide range of units for different geometric properties, some of which include the following (Yang et al., 2016):

1. **Length (Meter):** The meter (m) is the SI unit for measuring length. It is defined as the distance light travels in a vacuum during a specific fraction of a second. For example, the length of a line segment, a room's dimensions, or a building's height can all be measured in meters.

   For instance, if a line segment measures 5 meters, the equation would be:

   $$Length = 5 \times 1m = 5, meters$$

This equation illustrates the fundamental concept of measuring length in meters, where the numerical value represents how many times the standard unit (1 meter) fits into the given length.

2. **Area (Square Meter):** The square meter (m^2) is used to quantify the extent of two-dimensional surfaces. It is particularly relevant when measuring the area of geometric shapes like rectangles, triangles, and circles. The area of a rectangle is calculated by multiplying its length by its width, both measured in meters. For example,

   $$A = 5m\,x\,3m = 15\,m^2$$

3. **Volume (Cubic Meter):** The cubic meter (m^3) is the unit for measuring three-dimensional space. It is commonly employed when determining the volume of geometric solids, such as cubes, cylinders, and spheres. For instance, a cube's volume can be calculated using cubic meters.
4. **Angle (Radian):** While degrees are a common unit for measuring angles, the SI system recognizes the radian (rad) as a standard unit. A radian is the angle subtended at the center of a circle by an arc equal in length to the circle's radius. It is particularly useful in trigonometry and calculus (Bai et al., 2014).

1.4.2.2. Degrees and Minutes

While the SI system is widely adopted, degrees and minutes remain common for measuring angles in geometry, especially for everyday use. A degree (°) is further divided into 60 minutes ('). Each minute is divided into 60 seconds ("). This system allows for precise measurement of angles, with one full circle comprising 360°, each degree having 60 minutes and each minute containing 60 seconds (Jia et al., 2016).

1.4.2.3. Practical Application of Units

Units of measurement in geometry find extensive applications in various fields. Architects employ meters and square meters to design structures, ensuring accurate dimensions. In cartography, geographic measurements are crucial for creating precise maps. Astronomers use astronomical units (AU) to measure vast distances within the solar system, making calculations involving celestial bodies more manageable (DeGraff & Aydin, 1987).

1.4.2.4. Unit Conversions

In many cases, it becomes necessary to convert between different units of measurement. This process involves using conversion factors to express a measurement of another unit. For instance, converting from meters to centimeters involves multiplying by 100 since there are 100 centimeters in one meter. Such conversions are essential for maintaining consistency in geometric calculations. These equations demonstrate the conversion process between meters and centimeters, emphasizing the use of conversion factors to maintain consistency in geometric calculations.

Converting meters to centimeters:

$$Length\,in\,Centimeters = Length\,in\,Meters \times 100$$

For example, to convert $2\,meters$ to 2 centimeters (cm):

$$Length\,in\,Centimeters \;=\; 2\,m\,x100 = 200\,cm$$

1.4.3. Precision and Accuracy in Measurement

Precision and accuracy are fundamental concepts in measurement, playing a critical role in geometry and various scientific disciplines. This section will explore these concepts in depth, emphasizing their mathematical significance, applications, and the distinctions between them (Anderson et al., 2001).

1.4.3.1. Precision

Precision refers to the degree of consistency and reproducibility in measurements. In mathematical terms, it measures the scatter or spread of measurements around a central value. A high level of precision implies that repeated measurements of the same quantity will yield similar results.

Precision can be mathematically represented using the concept of standard deviation (σ). For a set of measurements {x1, x2, …, xn}, the standard deviation σ is calculated as:

$$\sigma = \sqrt{\Sigma\left(x_i - \bar{x}\right)^2 / (n-1)}$$

where; x_i represents individual measurements; (σ) is the mean (average) of the measurements; and n is the number of measurements. A smaller standard deviation indicates higher precision, as the measurements are clustered closely around the mean.

- **Example:** Consider measuring the sides of a square using a precise ruler. It demonstrates high precision if multiple measurements yield nearly identical values (e.g., 10.01 cm, 10.02 cm, 10.00 cm).

1.4.3.2. Accuracy

Accuracy, however, relates to how close a measured value is to a quantity's true or accepted value. In mathematical terms, accuracy quantifies the systematic error or bias in measurements. An accurate measurement has minimal systematic error and is close to the true value (Wylie, 1952).

Accuracy can be mathematically represented by the relative error (ε), given by:

$$\varepsilon = \left|\left(x - x_t\right)\right| / \left|x_t\right| * 100\%$$

where; x represents the measured value; and x� is the true or accepted value.

A smaller relative error indicates higher accuracy, signifying that the measured value is closer to the true value.

- **Example:** Suppose you are measuring the length of a standardized rod with a ruler known to have minimal systematic error. It indicates high accuracy if your measurement yields a value close to the known length (e.g., 25.00 cm measured against a 25.00 cm standard).

1.4.3.3. Precision vs. Accuracy

It is essential to distinguish between precision and accuracy, as they address different aspects of measurement. Here are some key points of differentiation (Jaswon & Dove, 1960):

1. **Precision Deals with Reproducibility:** Precision assesses how closely repeated measurements agree. It indicates the degree of scatter in measurements.
2. **Accuracy Pertains to Correctness:** Accuracy evaluates how close measured values are to the true or accepted values. It assesses systematic errors and biases in measurements.
3. **Both are Desirable:** A measurement should exhibit high precision and high accuracy. However, achieving both at the same time can be challenging.

4. **Trade-Off:** In practice, there is often a trade-off between precision and accuracy. Enhancing precision may require using more sophisticated equipment or techniques, which can introduce systematic errors and affect accuracy.

1.4.3.4. Applications in Geometry

Precision and accuracy are crucial when measuring lengths, angles, and other geometric properties in geometry. For example, in architectural design, precise measurements of building dimensions ensure that structures are built to specification. Surveyors rely on accurate measurements of land boundaries to create precise maps (Bai et al., 2014).

Hence, precision and accuracy are fundamental concepts in the measurement field, including geometry. While precision evaluates the consistency and scatter of measurements, accuracy assesses how close measured values are to the true values. Both are essential for reliable and meaningful measurements and understanding the trade-off between them is vital for achieving quality results in geometry and various scientific endeavors.

1.5. COORDINATE GEOMETRY

Coordinate geometry, often Cartesian geometry, is a powerful branch of mathematics that unites algebra and geometry. It systematically represents geometric shapes and explores their properties using algebraic equations and coordinates on a grid. At its core, coordinate geometry introduces the Cartesian coordinate system, where points in a plane are identified by their distances along two perpendicular axes (usually the x-axis and y-axis). This approach allows mathematicians and scientists to describe geometric objects, study relationships, and solve various problems, from graphing simple lines to analyzing complex curves and shapes. Coordinate geometry is a foundation of various fields, including physics, engineering, computer graphics, and more, making it an indispensable tool for understanding and working with spatial relationships (Rho et al., 2010).

- Example: Equation of a Line (Point-Slope Form)

The equation of a line with slope (m) passing through point (x_1, y_1) is given by:

$$y - y_1 = m(x - x_1)$$

This equation represents a line in the plane and is particularly useful in coordinate geometry for describing straight lines based on their slope and a specific point they pass through.

1.5.1. Cartesian Coordinate System

The Cartesian coordinate system, named after the renowned French mathematician René Descartes, is the basis of coordinate geometry. It provides a precise way to pinpoint locations in a two-dimensional plane using pairs of real numbers. This system introduces two perpendicular lines, typically the x-axis and y-axis, intersecting at a common point called the origin, usually denoted as (0, 0). These axes divide the plane into four quadrants: the first quadrant, where both x and y are positive; the second quadrant, where x is negative and y is positive; the third quadrant, with negative values for both x and y; and the fourth quadrant, where x is positive, and y is negative (Bodduluri & Ravani, 1993).

In the system, any point in the plane is uniquely identified by its coordinates, written as (x, y), where 'x' represents the horizontal distance along the x-axis and 'y' represents the vertical distance along the y-axis. The Cartesian coordinates enable us to describe various geometric shapes and objects, from simple points and lines to intricate curves and polygons. They form the foundation for various mathematical concepts and analytical tools, allowing us to precisely analyze equations, plot graphs, and solve geometric problems.

Let's delve deeper into the essential elements of the Cartesian coordinate system (Jia et al., 2016):

1. **The Origin (0, 0):** The origin serves as the reference point from which all other points in the plane are measured. It's where both the x and y coordinates are zero. In many applications, this point represents a starting point, a reference location, or a point of equilibrium.
2. **The x-Axis and y-Axis:** These two axes are straight lines that intersect at the origin, forming a right angle. The horizontal line is called the x-axis, and the vertical line is the y-axis. The x-axis extends infinitely to the right and left, while the y-axis extends infinitely upward and downward.
3. **Quadrants:** The plane is divided into four quadrants, designated as the first, second, third, and fourth quadrants. The location of

a point within these quadrants is determined by the signs of its x and y coordinates. Understanding which quadrant, a point lies in is crucial for many geometric and algebraic operations.

4. **Coordinates:** The coordinates of a point, written as (x, y), provide a numerical representation of its location. The 'x' coordinate tells us how far the point is horizontally from the origin, with positive values to the right and negative values to the left. The 'y' coordinate indicates the point's vertical position, with positive values above the origin and negative values below it.
5. **Distance and Slope:** The Cartesian coordinate system allows us to calculate the distance between two points using the distance formula and determine the slope of a line connecting two points. These calculations are fundamental in geometry and have applications in various fields, such as physics, engineering, and computer science.
6. **Equation of a Line**: In the Cartesian plane, lines can be represented by linear equations of the form y = mx + b, where 'm' is the slope of the line, and 'b' is the y-intercept. This equation provides a clear and concise way to describe lines and their properties.

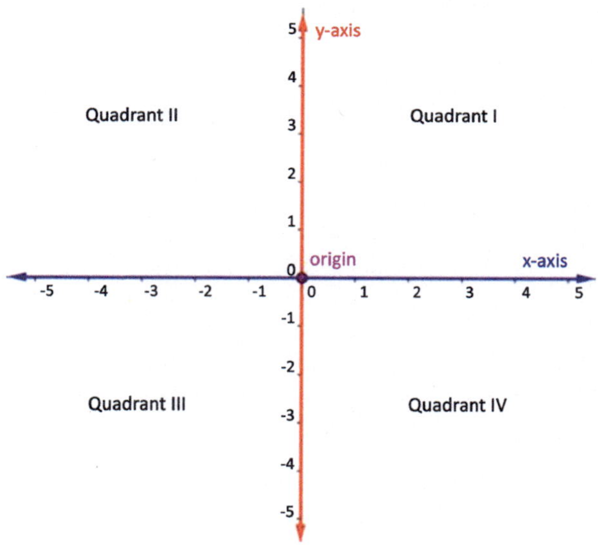

Figure 1.17. Illustration of a standard cartesian coordinate system.

Source: NROC, Creative Commons License.

7. **Graphing:** Plotting points and graphing equations are common in the Cartesian coordinate system. It allows us to visualize geometric shapes, explore mathematical relationships, and solve equations graphically (Figure 1.17).

Hence, the Cartesian coordinate system is a fundamental tool for studying geometry and analyzing geometric patterns mathematically.

1.5.2. Plotting Points and Lines on a Plane

In coordinate geometry, the ability to plot points and lines on a plane is fundamental. It lets us visually represent mathematical relationships, geometric shapes, and equations. This section will explore the step-by-step process of plotting points, drawing lines, and understanding the significance of these graphical representations within the Cartesian coordinate system (Snijders et al., 1993).

Remember

The Cartesian coordinate system is the foundation of coordinate geometry. It's a grid of horizontal and vertical lines where each point has a unique set of coordinates. This system simplifies the representation of complex shapes and equations in geometry.

1.5.2.1. Plotting Points

To plot a point on the Cartesian plane, we follow a straightforward procedure (Wylie, 1952):

1. **Understand the Coordinates:** Consider a point with coordinates (x, y). The 'x' coordinate tells us how far the point is horizontally from the origin, and the 'y' coordinate indicates its vertical position.
2. **Locate the Origin:** Identify the origin, typically denoted as (0, 0). This point serves as the reference for all other points.
3. **Horizontal Movement (x):** To locate the point along the x-axis, move horizontally from the origin. If 'x' is positive, move to the right; if 'x' is negative, move to the left. The magnitude of 'x' determines the distance.
4. **Vertical Movement (y):** Move vertically along the y-axis after positioning horizontally. If 'y' is positive, move upward; if 'y' is negative, move downward. Again, the magnitude of 'y' determines the distance.

5. **Intersection:** The point where your horizontal and vertical movements intersect is the plotted point. Label it with its coordinates (x, y).

- **Example:** Let's plot the point (–3, 2) on the Cartesian plane.
  - Begin at the origin (0, 0).
  - Move 3 units to the left (negative x).
  - Move 2 units upward (positive y).
  - The intersection of these movements represents the point (–3, 2) (Figure 1.18).

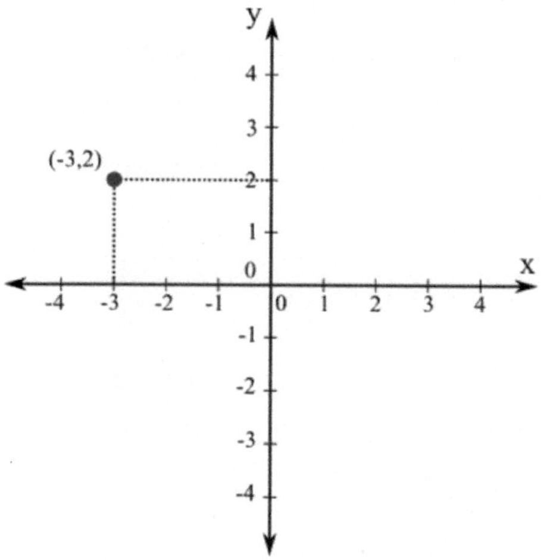

Figure 1.18. Location of a point (–3, 2) in a cartesian plane.

Source: Math Insight, Creative Commons License.

1.5.2.2. Drawing Lines

Lines in the Cartesian plane is often defined by linear equations in the form y = mx + b, where 'm' is the slope of the line, and 'b' is the y-intercept. Graphing lines is crucial in coordinate geometry (Vavra, 1990).

1. **Identify the Slope (m) and y-Intercept (b):** From the equation y = mx + b, determine the slope 'm' and the y-intercept 'b.' The slope represents how steep the line is, and the y-intercept is where the line crosses the y-axis.
2. **Plot the y-Intercept:** Begin by plotting the y-intercept (0, b) on the Cartesian plane.

3. **Use the Slope:** The slope 'm' indicates the rise (vertical change) over the run (horizontal change). For each unit you move horizontally, move 'm' units vertically based on the slope.
4. **Connect the Dots:** Draw a straight line using the plotted y-intercept and the slope. Extend it in both directions to cover the desired portion of the plane.

- **Example:** Graph the line y = 2x – 1.
  - Plot the y-intercept (0, –1).
  - Since the slope is 2, for every 1 unit to the right (horizontal), move 2 units up (vertical).
  - Connect the points to form a straight line (Figure 1.19).

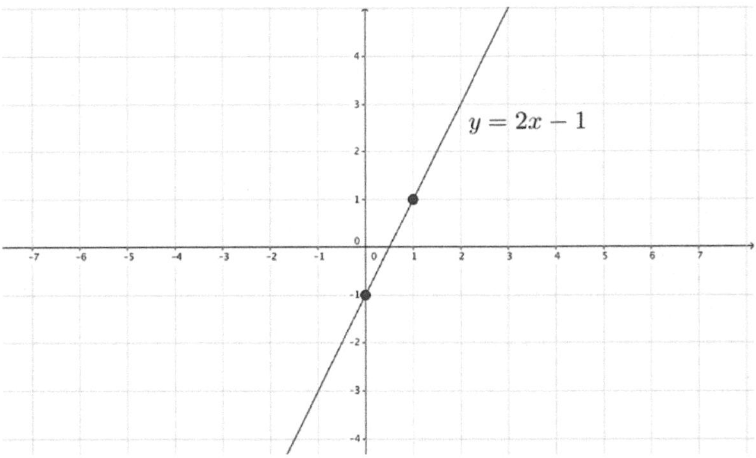

Figure 1.19. Drawing a line in the Cartesian coordinate system.

Source: Socratic, Creative Commons License.

1.5.2.3. Significance of Graphical Representations

Graphs in coordinate geometry provide a visual interpretation of mathematical relationships and concepts. They allow us to (Bai et al., 2014):

1. **Solve Equations Graphically:** Graphical representations help solve equations by finding the points where curves or lines intersect.
2. **Analyze Geometry:** Geometric shapes and figures can be visually understood and analyzed through their coordinates.

3. **Study Functions:** Functions and their behaviors are often visualized as curves or lines on the plane.
4. **Explore Real-World Applications:** Coordinate geometry finds applications in physics, engineering, economics, and various fields where spatial relationships and mathematical modeling are essential.
5. **Understand Trends:** By examining the slope and direction of lines, we can identify trends and make predictions based on data.

In summary, plotting points and drawing lines in the Cartesian coordinate system is a fundamental skill in geometry and mathematics. It serves as the basis for understanding more complex geometric concepts, equations, and spatial relationships, making it a valuable tool for theoretical and practical applications.

1.5.3. Distance and Midpoint Formulas

In coordinate geometry, two essential formulas are frequently used to calculate distances between points and find the midpoint of a line segment. These formulas are fundamental in various mathematical and practical applications, from analyzing geometric shapes to navigation and engineering. This section will explore the distance and midpoint formulas and understand their significance (Seidenberg, 1961).

1.5.3.1. Distance Formula

The distance between two points (x_1, y_1) and (x_2, y_2) in the Cartesian plane can be calculated using the distance formula derived from the Pythagorean theorem. The distance formula is as follows:

$$d = \sqrt{(x_2 - x_1)^2 + (y_2 - y_1)^2}$$

where; d represents the distance between the two points; x_1, y_1 are the coordinates of the first point; and x_2, y_2 are the coordinates of the second point.

- **Example:** Let's calculate the distance between points A (2, 3) and B (5, 7) (Figure 1.20).

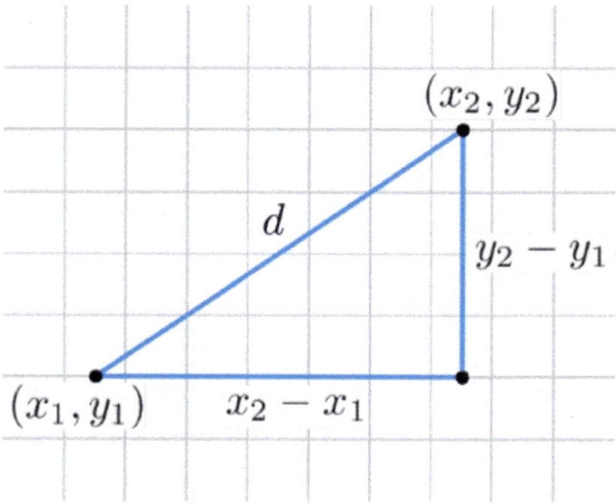

Figure 1.20. Distance formula to calculate distance between two points.

Source: Open Genus IQ, Creative Commons License.

Using the distance formula:

$$d = \sqrt{(5-2)^2 + (7-3)^2}$$

$$d = \sqrt{9+16}$$

$$d = \sqrt{25}$$

$$d = 5$$

So, the distance between points A and B is 5 units.

1.5.3.2. Midpoint Formula

The midpoint of a line segment is the point that divides the segment into two halves. To find the midpoint, we use the midpoint formula:

$$\mathrm{M}\left(\frac{x_1 + x_2}{2}, \frac{y_1 + y_2}{2}\right)$$

In this formula; M represents the midpoint; x_1, y_1 are the coordinates of the first endpoint; and x_2, y_2 are the coordinates of the second endpoint.

- **Example:** Let's find the midpoint of the line segment with endpoints C (2, 4) and D (8, 4) (Figure 1.21).

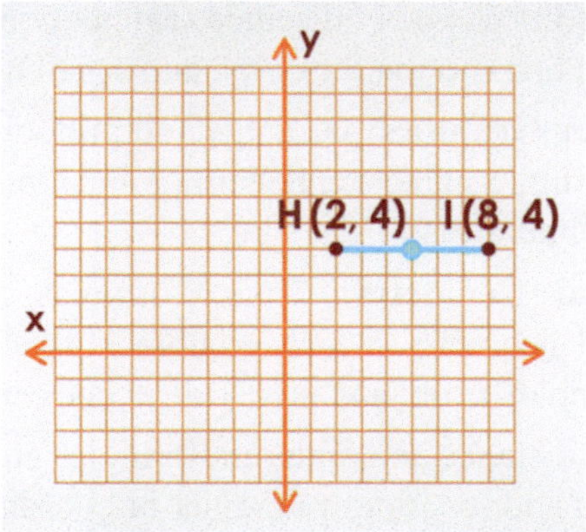

Figure 1.21. Schematic of calculation of midpoint.

Source: Tutors, Creative Commons License.

Using the midpoint formula:

$$M\left(\frac{2+8}{2},\frac{4+4}{2}\right)$$

$$M\left(\frac{10}{2},\frac{8}{2}\right)$$

$$M(5,4)$$

So, the midpoint of the line segment CD is M (5, 4).

1.5.3.3. Significance of Distance and Midpoint Formulas

These formulas play a crucial role in coordinate geometry and beyond (Ashraf et al., 2021):

1. **Geometric Analysis**: Distance formulas allow us to measure the lengths of line segments, sides of polygons, and distances between any two points in space.
2. **Navigation:** The distance formula calculates the shortest distance between two locations on the Earth's surface in navigation and GPS systems.
3. **Engineering:** Engineers use these formulas to design structures, determine lengths in architectural plans, and calculate dimensions in various applications.

4. **Physics:** In physics, distance calculations are essential for understanding motion, velocity, and acceleration.
5. **Computer Graphics:** Distance and midpoint formulas are vital for rendering graphics, calculating object positions, and creating realistic simulations.
6. **Analytical Geometry:** These formulas are foundational in analytical geometry, which combines algebra and geometry to solve complex problems involving shapes and coordinates.

In summary, the distance and midpoint formulas are invaluable tools for measuring distances and locating midpoints in coordinate geometry. They have practical applications in various fields, making them essential concepts for students and professionals alike.

1.6. GEOMETRIC FORMULA

Geometric formulas are mathematical expressions that quantify various aspects of shapes and spatial relationships.

These formulas encompass various geometric properties, from calculating the area, perimeter, and volume of different figures to determining angles, distances, and coordinates.

They provide precise methods for solving geometric problems and analyzing the characteristics of shapes. Whether it is finding the area of a triangle, the volume of a sphere, or the length of a diagonal in a rectangle, geometric formulas are the mathematical tools that enable us to explore and understand the intricacies of the geometric world (Davesne et al., 2021).

1.6.1. Perimeter and Area of Common Shapes

In geometry, measuring shapes is quintessential to understanding their properties and applications. Perimeter and area are two fundamental measurements that play a pivotal role in geometry.

Perimeter, often called the 'boundary length,' quantifies the total distance around a shape, outlining its outer boundary. On the other hand, area encapsulates the 'space enclosed' by a shape, signifying the region it occupies within a given space.

This section will explore the specifics of calculating the perimeter and area of two elementary geometric shapes: triangles and rectangles (Davesne et al., 2021).

1.6.1.1. Triangles

1. **The Perimeter of a Triangle:** It is the sum of the lengths of its three sides. For a triangle with sides of lengths 'a,' 'b,' and 'c,' the perimeter 'P' can be expressed as (Figure 1.22):

$$P=a+b+c$$

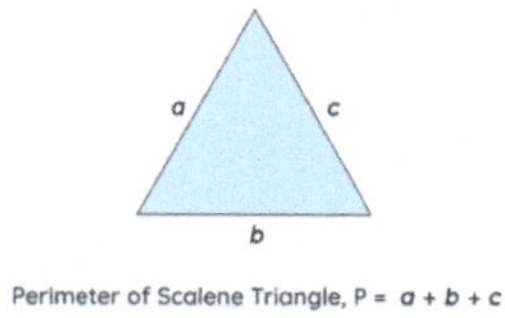

Figure 1.22. Illustration of the perimeter of a triangle.

Source: Cuemath, Creative Commons License.

This simple formula holds for all types of triangles, whether they are equilateral (with all sides of equal length), isosceles (with two sides of equal length), or scalene (with all sides of different lengths).

2. **Area of a Triangle:** The area 'A' of a triangle can be calculated using various methods, but one of the most common is by using the formula derived from its base 'b' and height 'h':

$$A=\frac{1}{2}\cdot b\cdot h$$

Here, 'b' represents the length of the triangle's base, and 'h' signifies the perpendicular distance from the base to the opposite vertex. The '1/2' factor is included to account for the fact that a triangle is half of a parallelogram with the same base and height. This formula applies to all types of triangles (Figure 1.23).

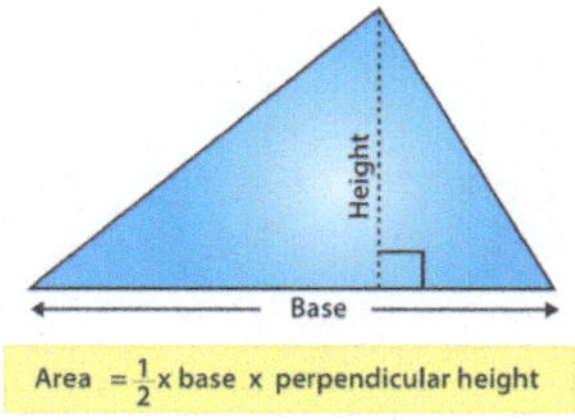

Figure 1.23. Schematic of an area of a triangle.

Source: BYJU's, Creative Commons License.

1.6.1.2. Rectangles

1. **The Perimeter of a Rectangle:** It is the sum of the lengths of all its sides. In a rectangle with length 'l' and width 'w,' the perimeter 'P' can be defined as:

$$P=2l+2w$$

Since rectangles have opposite sides of equal length, we can simplify this formula further to:

$$P=2(l+w)$$

This equation expresses that the perimeter is twice the sum of the length and width.

2. **Area of a Rectangle:** The area 'A' of a rectangle can be calculated using the formula:

$$A=l\cdot w$$

Here, 'l' represents the length of the rectangle, and 'w' signifies its width. This formula showcases the straightforward relationship between the length and width of a rectangle (Davesne et al., 2021) (Figure 1.24).

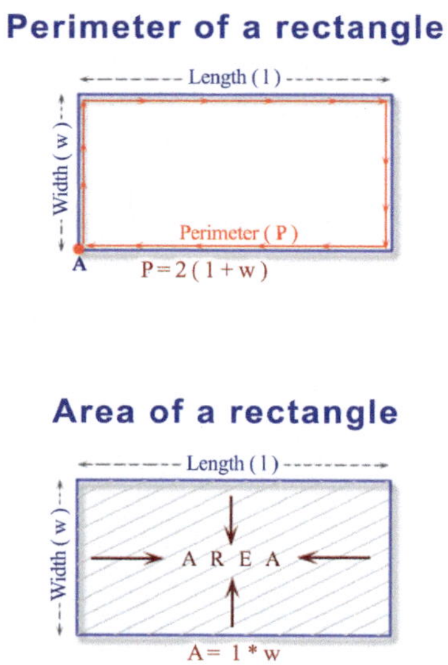

Figure 1.24. Perimeter and area of a rectangle.

Source: W3resource, Creative Commons License.

By understanding and applying these formulas, one can efficiently compute the perimeter and area of triangles and rectangles. These foundational principles lay the groundwork for more intricate geometric concepts and their real-world applications.

1.6.2. Volume and Surface Area of Three-Dimensional Shapes

Three-dimensional shapes introduce an additional layer of complexity compared to their two-dimensional counterparts. Understanding and calculating these shapes' volume and surface area are crucial for real-world applications. This section studies the concepts of volume and surface area applied to two fundamental three-dimensional shapes: cubes and spheres (Reiss et al., 2008).

1.6.2.1. Cubes

1. **Volume of a Cube:** A cube is a three-dimensional shape with all sides of equal length. To calculate its volume, we can use the formula:

$$V = a^3$$

where; 'V' represents the volume of the cube; and 'a' signifies the length of one side, also known as the edge length. This formula illustrates that the volume of a cube is obtained by cubing (taking power of 3) the length of any of its sides.

2. **Surface Area of a Cube:** A cube's surface area, 'SA,' is determined by adding up the areas of all its faces. Since a cube has six identical square faces, each with an area.' a^2,' we can calculate its surface area as:

$$SA = 6a^2$$

This formula reveals that the surface area of a cube is six times the square of its side length (Figure 1.25).

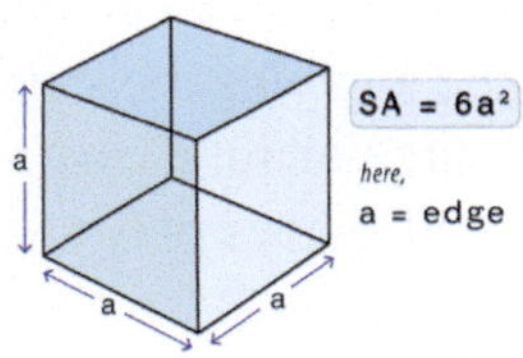

Figure 1.25. Schematic of the surface area of the cube.

Source: Math Monks, Creative Commons License.

1.6.2.2. Spheres

1. **Volume of a Sphere:** A sphere is a perfectly symmetrical three-dimensional shape with all points on its surface equidistant from its center. To find the volume of a sphere, we can use the formula:

$$V = \frac{4}{3} \eth r^3$$

where; 'V' represents the volume of the sphere; 'π' (pi) is approximately 3.14159; and 'r' denotes the sphere's radius. This formula demonstrates that the volume of a sphere is four-thirds of pi times the cube of its radius.

2. **Surface Area of a Sphere:** The surface area 'SA' of a sphere can be calculated using the formula:

$$SA = 4\pi r^2$$

Here; 'SA' represents the surface area; 'π' (pi) is approximately 3.14159; and 'r' signifies the sphere's radius. This formula highlights that the surface area of a sphere is four times pi times the square of its radius (Figure 1.26).

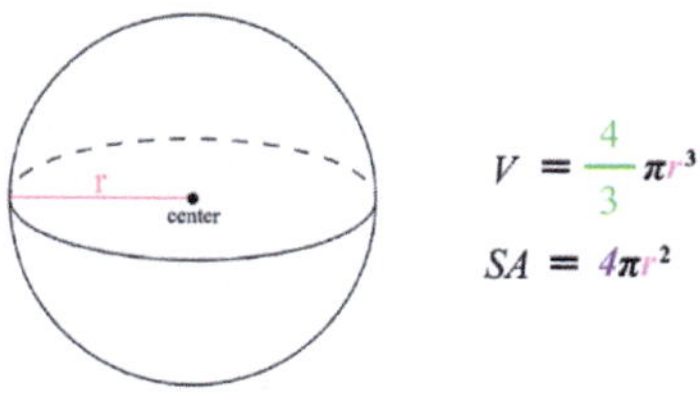

Figure 1.26. Schematic of the surface area of the sphere.

Source: Calcworkshop, Creative Commons License.

These formulas provide the mathematical foundation for calculating cubes and spheres' volume and surface area, two of the most fundamental three-dimensional shapes. We can solve geometric problems and practical scenarios by understanding and applying these principles. These calculations are pivotal in fields such as architecture, engineering, physics, and many others, where three-dimensional shapes play a significant role.

1.7. MATHEMATICAL REASONING IN GEOMETRY

Mathematical reasoning is the backbone of geometry, serving as the driving force behind the exploration and understanding of spatial relationships and shapes. In geometry, we employ deductive reasoning to formulate logical arguments and proofs. The deductive reasoning in geometry explains that the process employes logic to draw conclusions based on formerly recognized premises or facts. As deductions are based on earlier known proofs, it is more appropriate to reach right conclusions when compared to inductive reasoning.

Starting with axioms, definitions, and previously established theorems, we construct step-by-step arguments to demonstrate the validity of geometric statements. This reasoning process is crucial in establishing the properties of various shapes, angles, and figures. It enhances our ability to solve geometric problems and sharpens our analytical thinking skills, enabling us to apply the principles of geometry to real-world scenarios. Whether it is proving the congruence of triangles or determining the symmetry of a complex polyhedron, mathematical reasoning forms the solid foundation upon which geometric knowledge is built, making it an essential skill in studying geometry (Eisenhart, 2015).

1.7.1. Deductive Reasoning

In geometry, deductive reasoning is a fundamental method of drawing logical conclusions based on established principles and previously proven statements. It relies on axioms, definitions, and theorems to deduce new truths. Starting with known facts, deductive reasoning allows mathematicians to systematically apply logical steps to reach precise conclusions about geometric shapes, angles, and spatial relationships. By using deductive reasoning, mathematicians can construct rigorous proofs, ensuring the validity of geometric statements and expanding our understanding of the intricate patterns and properties within the world of geometry (Mathai, 1999).

1.7.1.1. The Role of Axioms and Definitions

Deductive reasoning in geometry begins with axioms and definitions. Axioms, or postulates, are self-evident truths or assumptions that form the foundation of geometric reasoning. These statements are considered so fundamental that they do not require proof. For example, one common axiom in Euclidean geometry is the statement that through two distinct points, there is exactly one straight line (Trofimov, 2013).

On the other hand, definitions clarify and specify the meaning of geometric terms. They are essential for establishing a common language and understanding among mathematicians. For instance, the definition of a circle provides a precise description of its geometric shape: "A circle is the set of all points in a plane that are equidistant from a given point called the center."

1.7.1.2. Logical Steps in Deductive Reasoning

We can proceed with deductive reasoning once we have our axioms and definitions. The process typically involves logical steps supported by prior statements or theorems. Let's explore a simple example to illustrate deductive reasoning in action (Minsky & Papert, 1969):

- **Statement 1:** Given two points, A and B, a unique line l passes through both points (axiom).
- **Statement 2:** Points A and B are given (given).
- **Statement 3:** Line l passes through points A and B (Statement 1).
- **Statement 4:** Therefore, points A and B lie on line l (definition of a line).

In this example, we use deductive reasoning to conclude that points A and B lie on the line l. This process involves linking each statement logically to the preceding ones, ultimately leading to the desired conclusion.

1.7.1.3. Formal Proofs and Theorems

Deductive reasoning in geometry often culminates in formal proofs of theorems. A theorem is a statement proven true based on a logical sequence of deductive steps. These proofs serve as irrefutable evidence of the validity of geometric principles.

For instance, one classic theorem in geometry is the Pythagorean theorem: "The Pythagorean Theorem: In a right triangle, the square of the length of the hypotenuse (the side opposite the right angle) is equal to the sum of the squares of the lengths of the other two sides."

The deductive proof of the Pythagorean Theorem involves constructing squares on each side of the right triangle, demonstrating that the areas of these squares obey the theorem's relationship.

1.7.1.4. Applications in Geometric Problem Solving

Deductive reasoning is an essential tool for solving geometric problems. Whether proving that a quadrilateral is a parallelogram, determining the congruence of two triangles, or establishing the properties of a circle's tangents, mathematicians rely on deductive methods to arrive at conclusive solutions. For example, the problem of proving that opposite angles of a parallelogram are congruent:

1. **Problem:** Given a parallelogram ABCD, prove that angle A is congruent to angle C, and angle B is congruent to angle D (Figure 1.27).

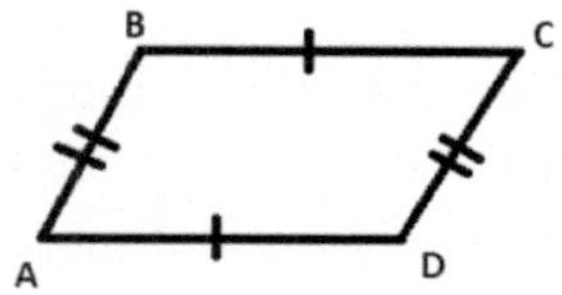

Figure 1.27. Schematic of a parallelogram.

Source: Study, Creative Commons License.

- **Proof:**
  - **Statement 1:** Parallelogram ABCD is given (given).
  - **Statement 2:** AB is parallel to CD, and AD is parallel to BC (definition of a parallelogram).
  - **Statement 3:** Line AD intersects line BC, forming transversals AB and CD (Statement 2).
  - **Statement 4:** Angle A and angle C, are corresponding angles formed by transversals AB and CD (definition of corresponding angles).
  - **Statement 5:** Therefore, angle A is congruent to angle C (Definition of congruent angles).

A similar argument can be made to prove the congruence of angles B and D. This concise proof relies on deductive reasoning to establish the desired congruence relationships.

1.7.2. Logical Arguments and Proofs

Logical arguments and proofs are the basis of mathematical reasoning in geometry. This section will study geometry's precise methodologies and techniques to construct logical arguments and establish proofs of geometric truths.

1.7.2.1. The Structure of a Proof

Proof in geometry follows a structured format that is both rigorous and logical. It typically consists of several key components:

- **Statement of the Theorem:** A clear and concise statement of the theorem or proposition to be proven. It sets the goal for the proof.

1. **List of Given Information:** Any relevant information or conditions in the problem statement. These are the initial building blocks of the proof.
2. **Construction:** If necessary, construct additional points, lines, or shapes that aid the proof. Constructed elements are introduced to simplify the problem and make it more approachable.
3. **Logical Deductions:** A series of logical steps and deductions that lead from the given information and constructions to the desired conclusion. Each step must be justified based on prior statements or accepted axioms and theorems.
4. **Conclusion:** The final statement summarizes the proof and confirms the theorem's validity.

- **Example:** Proof of the Triangle Sum Theorem

  Let's illustrate these components by proving the Triangle Sum Theorem, which states that the sum of the interior angles of a triangle is always 180°.

  Theorem (Triangle Sum Theorem): In any triangle, the sum of its three interior angles equals 180°.

- **Given:** Consider triangle ABC.
- **Construction:** Extend side AB to form a line segment BD (Figure 1.28).

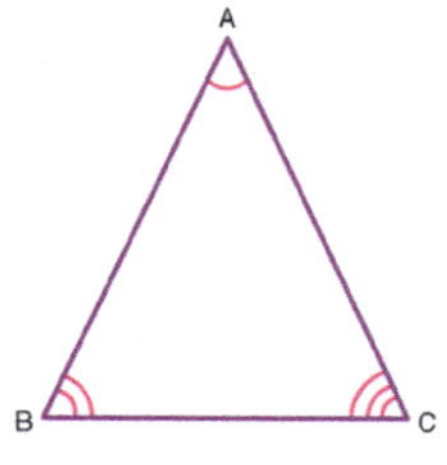

Figure 1.28. Illustration of a triangle ABC.

Source: BYJU's, Creative Commons License.

- **Logical Deductions:**

1. Angle A and angle D are vertical angles, therefore, congruent (Vertical angles are congruent by definition).
2. Angle A and angle B form a straight line, so they are supplementary (Definition of supplementary angles).
3. Angle D and angle C form a straight line, so they are supplementary (Definition of supplementary angles).
4. The sum of angles A and B equals the sum of angles B and D, which are supplementary (Transitive property of equality).
5. Therefore, the sum of angles A, B, and C equals 180° (Figure 1.29).

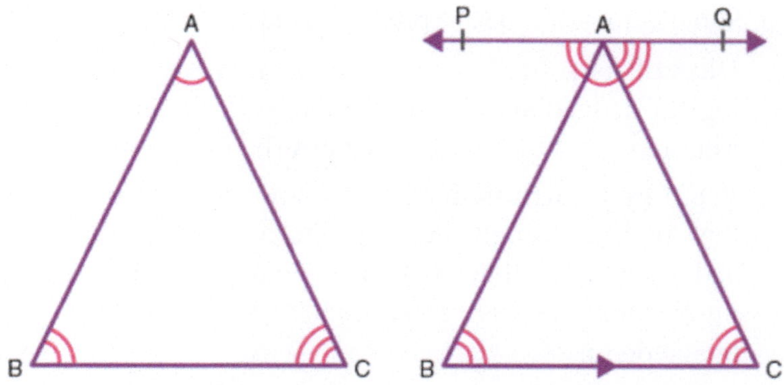

Figure 1.29. Triangle showing different angle combinations.

Source: BYJU'S, Creative Commons License.

- **Proof:**

  PQ is a straight line which states that:

  $\angle QAC + \angle PAB + \angle BAC = 180°$ (1)

  Since BC||PQ and AC, AB are transversals,

  So, $\angle QAC = \angle ACB$ (alternate angle pair)

  Also, $\angle PAB = \angle CBA$ (alternate angle pair)

  Substituting values of $\angle PAB$ and $\angle QAC$ in Eqn. (1):

  $\angle ACB + \angle CBA + \angle BAC = 180°$

- **Conclusion:** The sum of the interior angles of triangle ABC is 180°, confirming the Triangle Sum Theorem.

1.7.2.2. The Role of Axioms and Postulates

Axioms and postulates are the starting points from which all other geometric facts are derived. While axioms are self-evident factual statements, postulates are specific statements that describe geometric properties. For example, Euclid's Fifth Postulate, also known as the Parallel Postulate, states that if a straight line falls on two straight lines in such a way that the interior angles on the same side are less than two right angles, then the two lines, if extended indefinitely, will meet on that side. This postulate and other axioms form the foundation of Euclidean geometry.

1.7.2.3. Types of Proofs

In geometry, several methods of proof are employed, each with its unique approach. Some common types of proofs include:

1. **Direct Proof:** In direct proof, a theorem is proven by a series of logical deductions that lead directly to the desired conclusion. The steps are typically laid out clearly and linearly.
2. **Proof by Contradiction:** Also known as reductio ad absurdum, this method assumes the opposite of what needs to be proven and shows that it leads to a contradiction. This contradiction implies that the original assumption was false, thus establishing the theorem.
3. **Proof by Mathematical Induction:** This technique often proves statements about integers or sequences. It involves proving that a statement holds for a base case and then showing that it must also hold for the next case if it holds for any given case.

- **Example:** Proof by Contradiction

Show that if two angles in a triangle are congruent, the sides that face those angles are also congruent.

- **Proof (By Contradiction):**
  - **Assumption:** Consider a triangle with two congruent angles but non-congruent sides on either side of those angles. One of the sides must be longer if the sides are not congruent (Figure 1.30).

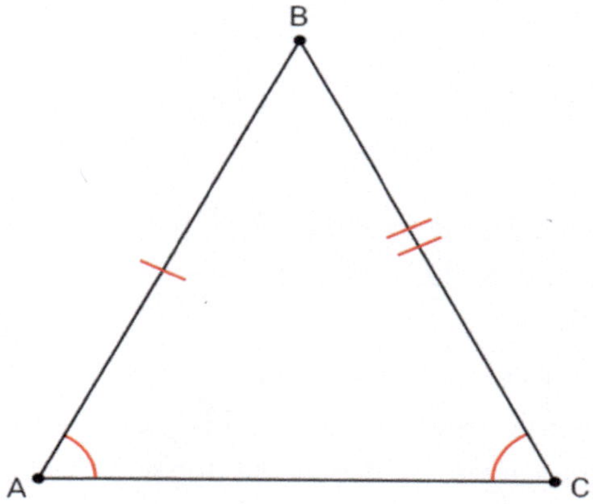

Figure 1.30. Triangle with two congruent angles.

Source: Brilliant.org, Creative Commons License.

In the above triangle, ∠A≅∠C, and AB>BC.

Since, $\overline{BC}$ is shorter than $\overline{AB}$, place D on $\overline{AB}$ such that AD=BC (Figure 1.31).

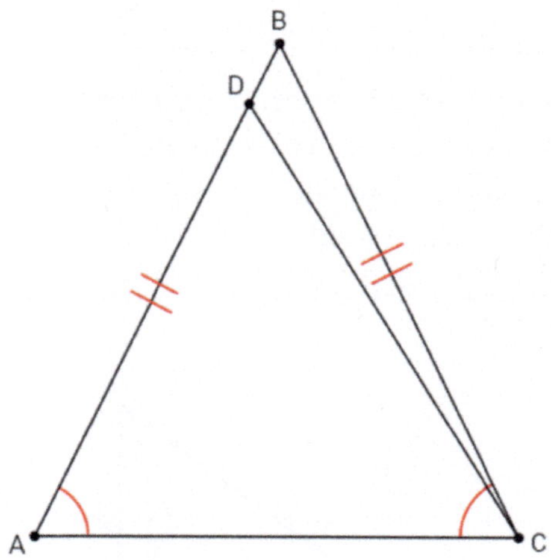

Figure 1.31. Modified triangle for proof of contradiction.

Source: Brilliant.org, Creative Commons License.

Now the following congruences can be stated:

$\overline{AD} \cong \overline{BC}$

$\angle A \cong \angle ACB$

$\overline{AC} \cong \overline{AC}$

- **Conclusion:** These congruences imply the triangular congruence of ΔABC and ΔCDA. This, however, is illogical because ΔABC is evidently larger.

Hence, logical arguments and proofs in geometry are rigorous and systematic processes that establish the validity of geometric theorems and propositions. By adhering to a structured format, relying on axioms and postulates, and employing various proof techniques, mathematicians ensure the precision and accuracy of geometric reasoning, enabling the exploration of the rich and complex world of geometry.

1.7.3. Example of Geometric Proof

In this section, we will explore a selection of geometric proofs to illustrate the application of logical arguments and deductive reasoning in geometry. These proofs are concrete examples of how mathematical reasoning is applied to establish geometric truths.

- Proof of the Pythagorean Theorem:
  - **Theorem (Pythagorean Theorem):** In a right triangle, the square of the length of the hypotenuse (the side opposite the right angle) is equal to the sum of the squares of the lengths of the other two sides.
- **Given:** Consider a right triangle ABC with a right angle at C (Figure 1.32).

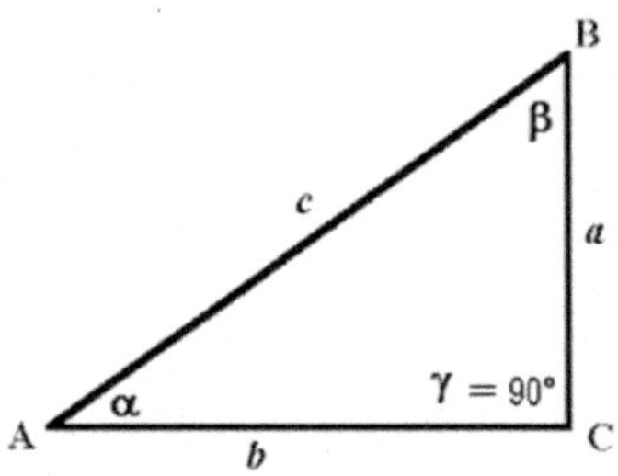

Figure 1.32. Triangle ABC with a right angle at C.

Source: Jorge M. López Fernández, Creative Commons License.

- **Construction:** Drop a perpendicular from vertex C to side AB, creating two right triangles, ADC and BDC.
- **Logical Deductions:**

1. In triangle ADC, by definition of cosine, we have:

$$\cos(\Box ADC) = \frac{AD}{AC}$$
$$AD = AC \cdot \cos(\Box ADC)$$

2. In triangle BDC, by definition of sine, we have:

$$\sin(\Box BDC) = \frac{BD}{BC}$$

$$BD = BC \cdot \text{ü} \ (\Box BDC)$$

3. Since ADC and BDC are complementary angles,

$$\Box ADC + \Box BDC = 90^\circ$$

4. Using trigonometric identities, we have:

$$\cos(\Box ADC) = \sin(\Box BDC)$$

5. Combining Eqns. (1) and (2) with (4), we get:

$$AD = AC \cdot \sin(\Box BDC)$$

6. In triangle ABC, by definition of sine, we have:

$$\sin(\Box BDC) = \frac{BD}{AB}$$

$$BD = AB \cdot \sin(\Box BDC)$$

7. Substituting Eqn. (5) into Eqn. (6), we obtain:

$$AD = AC \cdot \sin(\Box BDC) = AB \cdot \sin(\Box BDC)$$

8. Since AD = BD, we can equate the expressions:

$$AC = AB$$

Squaring both sides:

$$AC^2 = (AB)^2$$

2. From the original triangle ABC, by the Pythagorean Theorem, we have:

$$AC^2 + BC^2 = AB^2$$

10. Substituting Eqn. (9) into Eqn. (10), we obtain:

$$AB^2 + BC^2 = (AB)^2$$

11. Canceling out AB^2 on both sides:

$BC^2 = (AB)^2 - AC^2$

12. Substituting Eqn. (9) into Eqn. (12), we get:

$BC^2 = (AC)^2 - AC^2$

13. Simplifying:

$BC^2 = AC^2 - AC^2$

$BC^2 = 0$

14. Taking the square root of both sides:

$BC = 0$

15. Since BC cannot be zero (it is a side length), our original assumption of AD = BD was incorrect. Therefore, we conclude that AD ≠ BD.
16. Therefore, AD + BD = AB.
17. Substituting AD and BD with their respective expressions from Eqns. (5) and (6):

$AC \cdot \sin(\square BDC) + AB \cdot \sin(\square BDC) = AB$

18. Factoring out AB and using the distributive property:

$AB \cdot (\sin(\square BDC) + \sin(\square BDC)) = AB$

19. Combining like terms:

$2AB \cdot \sin(\square BDC) = AB$

20. Dividing both sides by 2AB:

$\sin(\square BDC) = \frac{1}{2}$

21. In triangle BDC, by definition of sine:

$\sin(\square BDC) = \frac{BD}{BC}$

22. Substituting the value of $\sin(\square BDC)$ *from* Eqn. (21):

$\frac{1}{2} = \frac{BD}{BC}$

23. Cross-multiplying:

$BD = \frac{1}{2} \cdot BC$

24. Substituting BD with its expression from Eqn. (2):

$$\frac{1}{2} \cdot BC = BC \cdot \text{ü} \ (\square BDC)$$

25. Simplifying:

$$BC \cdot \sin(\square BDC) = \frac{1}{2} \cdot BC$$

1.8. REAL-WORLD APPLICATIONS OF GEOMETRY

Geometry is not just an abstract mathematical discipline but a powerful tool with many real-world applications, some of which are pivotal in architecture and engineering. In these domains, geometric concepts and principles serve as the foundation for creating structures that are aesthetically pleasing and structurally sound (Barneva & Brimkov, 2009).

1.8.1. Architectural Design

Geometry is fundamental to architectural design. Architects use geometric principles to create structurally sound and visually appealing buildings. Concepts like symmetry, proportion, and balance, all rooted in geometry, guide the layout and aesthetics of architectural designs. For instance, using the Golden Ratio, a geometric proportion, can lead to harmonious and visually pleasing buildings (Figure 1.33).

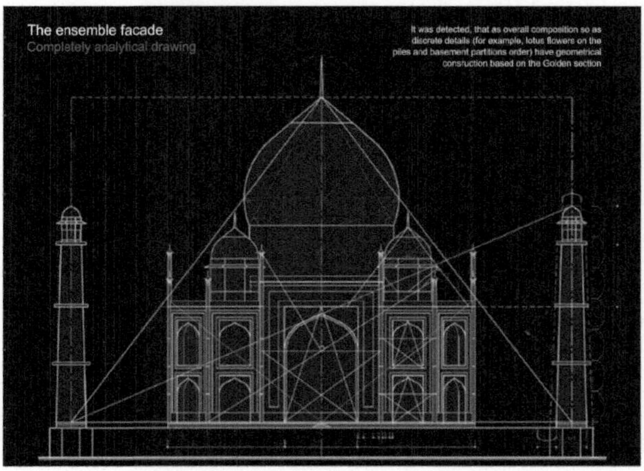

Figure 1.33. The world-renowned monument displays a golden ratio in the width of its facade with a grand central arch and the heights of the windows.

Source: Tanvi Saraf, Creative Commons License.

Moreover, architects rely on geometric calculations to ensure the stability of structures. They use trigonometry to calculate angles, distances, and load-bearing capacities. For example, when designing a bridge, understanding the geometry of triangles and the distribution of forces is essential to ensure the bridge's structural integrity. Geometry provides architects with the tools to bring their creative visions to life while adhering to essential principles of stability and safety (Oke, 1981).

1.8.2. Engineering and Construction

Engineers harness the power of geometry to design and construct various structures, from bridges and tunnels to skyscrapers and dams. The geometric concept of vectors is crucial for analyzing forces and movements in structural engineering. Engineers use vectors to understand how forces act on buildings and infrastructure, helping them determine the materials needed and how they should be arranged (Rian & Asayama, 2016).

Geometry also plays a central role in surveying, a critical aspect of engineering and construction. Surveyors use principles of geometry to measure land, establish property boundaries, and create topographical maps. Tools like theodolites and total stations enable precise angle and distance measurements, facilitating accurate mapping and construction.

1.8.3. Geometric Modeling

In modern architecture and engineering, geometric modeling is a powerful technique. It involves creating three-dimensional digital models of structures using computer software. These models aid in visualizing the final product and provide valuable insights into structural integrity (Catterall et al., 1999).

For instance, in the design of an aircraft, geometric modeling allows engineers to simulate airflow over its surfaces, helping optimize its aerodynamics. Similarly, in the automotive industry, geometric modeling assists in designing vehicles with enhanced safety features and fuel efficiency.

1.8.4. Geometry in Environmental Planning

Environmental planning involves optimizing land use for various purposes, including residential, commercial, and recreational. Geometric principles are pivotal in creating functional and sustainable urban spaces (Catterall et al., 1999).

One example is the layout of road networks in a city. Engineers and urban planners use geometric concepts to design road systems that efficiently connect various parts of the city, minimizing traffic congestion and travel times. Similarly, arranging parks and green spaces in an urban environment often follows geometric patterns to ensure accessibility and aesthetics.

1.8.5. Structural Analysis

Before any construction project begins, engineers must assess structures' stability and load-bearing capacity. It involves analyzing how different materials and components interact under various conditions (Hwang et al., 2007).

Geometry is instrumental in structural analysis. Engineers use mathematical equations to calculate building materials' stresses, strains, and deflections. By understanding the geometric properties of these materials, engineers can predict how they will behave when subjected to different forces.

1.8.6. Civil Engineering

In civil engineering, geometry is vital in projects related to water management, such as designing canals, dams, and sewage systems. Engineers use geometric calculations to determine optimal gradients, flow rates, and volumes in these systems. It ensures efficient and effective management of water resources, reducing the risk of flooding and providing access to clean water (Hwang et al., 2007).

1.8.7. Materials Science

Geometry also plays a role in materials science, where researchers study substances' atomic and molecular arrangement. The arrangement of atoms and molecules in a material can significantly impact its properties, such as strength, conductivity, and transparency.

Did You Know?

Geometry is essential in materials science. Researchers use geometric principles to study and design new materials, like advanced polymers and composites. The arrangement of atoms and molecules in these materials plays a critical role in their properties and applications.

By understanding the geometric structure of materials, scientists can develop new materials with enhanced properties for various applications, including electronics, aerospace, and healthcare (Hughes et al., 2005).

Geometry is not confined to textbooks and classrooms; it is an essential tool in real-world applications, particularly architecture, engineering, and environmental planning. From designing iconic skyscrapers to optimizing transportation networks, geometry provides the mathematical framework for turning creative ideas into practical, functional, and safe structures and systems. Its enduring significance in these fields underscores its vital role in shaping the modern world.

Activity 1.1

You are part of a team responsible for planning a sustainable urban development project. How can you apply geometric concepts and principles to design green spaces, optimize energy-efficient building layouts, and minimize the project's environmental impact?

1.9. SUMMARY

This chapter provides a fundamental introduction to geometry. It delves into the historical development of geometry and highlights its cultural importance. Basic geometric shapes like points, lines, planes, angles, and triangles are explained. Geometric constructions, including their practical applications, are covered. Measurement in geometry, precision, and units of measurement are discussed. Coordinate geometry and its role in plotting points and distances on a plane are introduced. Geometric formulas for calculating perimeter, area, volume, and surface area are presented. The chapter also explores mathematical reasoning and proofs in geometry. Finally, it touches upon the real-world applications of geometry in fields like architecture, engineering, and materials science.

REFERENCES

1. Aaron, W. R., & Herbst, P. G., (2019). The teacher's perspective on the separation between conjecturing and proving in high school geometry classrooms. *Journal of Mathematics Teacher Education*, *22*, 231–256.
2. Adams, D. C., & Otárola-Castillo, E., (2013). Geomorph: An R package for the collection and analysis of geometric morphometric shape data. *Methods in Ecology and Evolution*, *4*(4), 393–399.
3. Anderson, R. C., Dohm, J. M., Golombek, M. P., Haldemann, A. F., Franklin, B. J., Tanaka, K. L., & Peer, B., (2001). Primary centers and secondary concentrations of tectonic activity through time in the western hemisphere of Mars. *Journal of Geophysical Research: Planets*, *106*(E9), 20563–20585.
4. Ashraf, Q. H., Galor, O., & Klemp, M., (2021). The ancient origins of the wealth of nations. In: *The Handbook of Historical Economics* (Vol. 2, pp. 675–717). Academic Press.
5. Bai, G. Z., Liu, S. Y., Wang, H., Zhi, W., & Yin, Z. J., (2014). A novel method for smooth blending cylindrical surfaces whose axes is non-coplanar with intersecting line of two tapered surfaces. *Applied Mechanics and Materials*, *687*, 1470–1473.
6. Barka, A. A., & Kadinsky-Cade, K., (1988). Strike-slip fault geometry in Turkey and its influence on earthquake activity. *Tectonics*, *7*(3), 663–684.
7. Barneva, R. P., & Brimkov, V. E., (2009). Digital geometry and its applications to medical imaging. In: *Advances in Computational Vision and Medical Image Processing: Methods and Applications* (Vol. 2, pp. 77–92). Dordrecht: Springer Netherlands.
8. Barreto, J. P., & Araujo, H., (2005). Geometric properties of central catadioptric line images and their application in calibration. *IEEE Transactions on Pattern Analysis and Machine Intelligence*, *27*(8), 1327–1333.
9. Battista, M. T., (1990). Spatial visualization and gender differences in high school geometry. *Journal for Research in Mathematics Education*, *21*(1), 47–60.
10. Bennell, J. A., & Oliveira, J. F., (2008). The geometry of nesting problems: A tutorial. *European Journal of Operational Research*, *184*(2), 397–415.

11. Blanton, M. L., & Kaput, J. J., (2005). Characterizing a classroom practice that promotes algebraic reasoning. *Journal for Research in Mathematics Education*, *36*(5), 412–446.
12. Bodduluri, R. M. C., & Ravani, B., (1993). Design of developable surfaces using duality between plane and point geometries. *Computer-Aided Design*, *25*(10), 621–632.
13. Bozkurt, G., Uygan, C., & Turgut, M., (2018). In: Hans-Georg, W., Alison, C.-W., & Ana, D., (eds.), *Instrumental Genesis of a Preservice Mathematics Teacher: Instrumented Actions for Perpendicular Line Construction in a Dynamic Geometry Environment* (Vol. 43, pp. 9871–9875).
14. Bridge, G., (1997). Mapping the terrain of time—space compression: Power networks in everyday life. *Environment and Planning D: Society and Space*, *15*(5), 611–626.
15. Brodmann, M. P., & Sharp, R. Y., (2012). *Local cohomology: An Algebraic Introduction with Geometric Applications*. (Vol. 136, pp. 654–656). Cambridge University Press.
16. Bruce, C. D., Davis, B., Sinclair, N., McGarvey, L., Hallowell, D., Drefs, M., & Woolcott, G., (2017). Understanding gaps in research networks: Using "spatial reasoning" as a window into the importance of networked educational research. *Educational Studies in Mathematics*, *95*, 143–161.
17. Burger, W. F., & Shaughnessy, J. M., (1986). Characterizing the van Hiele levels of development in geometry. *Journal for Research in Mathematics Education*, *17*(1), 31–48.
18. Cai, Y. C., Wu, J., & Atluri, S. N., (2013). A new implementation of the numerical manifold method (NMM) for the modeling of non-collinear and intersecting cracks. *Computer Modeling in Engineering & Sciences*, *92*(1), 63–85.
19. Catterall, J. S., Chapleau, R., & Iwanaga, J., (1999). Involvement in the arts and human development: General involvement and intensive involvement in music and theater arts. *Champions of Change: The Impact of the Arts on Learning*, *1*, 1–18.
20. Cerda, E., & Mahadevan, L., (2003). Geometry and physics of wrinkling. *Physical Review Letters*, *90*(7), 074302.
21. Chandrasekaran, V., Recht, B., Parrilo, P. A., & Willsky, A. S., (2012). The convex geometry of linear inverse problems. *Foundations of Computational Mathematics*, *12*, 805–849.

22. Charalambos, L., (1997). A few remarks regarding teaching geometry through a theoretical analysis of the geometrical figure. *Nonlinear Analysis, Theory, Methods, and Applications*, *30*(4), 2087–2095.
23. Chin, H., & Chew, C. M., (2023). Expert-based cognitive models vs theory-based cognitive models: Which cognitive models match students' cognition on 'parallel and perpendicular lines' better? *Contemporary Educational Psychology*, *73*, 102175.
24. Cosgrove, D., (1985). Prospect, perspective, and the evolution of the landscape idea. *Transactions of the Institute of British Geographers*, *4*(4), 45–62.
25. Davesne, D., Friedman, M., Schmitt, A. D., Fernandez, V., Carnevale, G., Ahlberg, P. E., & Benson, R. B., (2021). Fossilized cell structures identify an ancient origin for the teleost whole-genome duplication. *Proceedings of the National Academy of Sciences*, *118*(30), 2101780118.
26. DeGraff, J. M., & Aydin, A., (1987). The surface morphology of columnar joints is significant to the mechanics and direction of joint growth. *Geological Society of America Bulletin*, *99*(5), 605–617.
27. Domski, M., (2004). Lorenzo Magnani, philosophy, and geometry: Theoretical and historical issues. The Western Ontario series in the philosophy of geometry (Vol. 66). New York: Kluwer Academic Publishers (2001), xix+ 249 p. 44. *Philosophy of Science*, *71*(3), 412–415.
28. Eisenhart, L. P., (2015). *Introduction to Differential Geometry* (Vol. 2227). Princeton University Press.
29. Eisenstein, E. L., (1969). The advent of printing and the problem of the renaissance. *Past and Present*, *45*, 19–89.
30. Fredriksson, R., Hägglund, M., Olszewski, P. K., Stephansson, O., Jacobsson, J. A., Olszewska, A. M., & Schiöth, H. B., (2008). The obesity gene, FTO, is of ancient origin, up-regulated during food deprivation, and expressed in neurons of feeding-related nuclei of the brain. *Endocrinology*, *149*(5), 2062–2071.
31. Gangnard, S., Chêne, A., Dechavanne, S., Srivastava, A., Avril, M., Smith, J. D., & Gamain, B., (2019). VAR2CSA binding phenotype has ancient origin and arose before *Plasmodium falciparum* crossed to humans: Implications in placental malaria vaccine design. *Scientific Reports*, *9*(1), 16978.

32. Gecü, Z., & Özdener, N., (2010). The effects of using geometry software supported by digital daily life photographs on geometry learning. *Procedia-Social and Behavioral Sciences*, *2*(2), 2824–2828.

33. Gibbons, T. H., (1981). Cubism and the fourth dimension in the context of the late 19th-century and early 20th-century revival of occult idealism. *Journal of the Warburg and Courtauld Institutes*, *44*(1), 130–147.

34. Gimbel, S., (2001). Flatland, curved space: How MC Escher illustrated the history of geometry. In: *Bridges: Mathematical Connections in Art, Music, and Science* (Vol. 2, pp. 123–132).

35. González, G., & Herbst, P. G., (2006). Competing arguments for the geometry course: Why were American high school students supposed to study geometry in the 20th century? *International Journal for the History of Mathematics Education*, *1*(1), 654–656.

36. Gulikers, I., & Blom, K., (2001). A historical angle,' a survey of recent literature on the use and value of history in geometrical education. *Educational Studies in Mathematics*, *47*, 223–258.

37. Habre, S., (2017). Students' challenges with polar functions: Covariational reasoning and plotting in the polar coordinate system. *International Journal of Mathematical Education in Science and Technology*, *48*(1), 48–66.

38. Herbst, P. G., (2002). Establishing a custom of proving in American school geometry: Evolution of the two-column proof in the early 20th century. *Educational Studies in Mathematics*, *49*, 283–312.

39. Hersh, R., (1979). Some proposals for reviving the philosophy of mathematics. *Advances in Mathematics*, *31*(1), 31–50.

40. Hillier, B., (1999). The hidden geometry of deformed grids: or, why space syntax works when it looks as though it shouldn't. *Environment and Planning B: Planning and Design*, *26*(2), 169–191.

41. Ho, S., Carruthers, D., & Imbert, P., (2016). Insights into the permeability of polygonal faults from their intersection geometries with linear chimneys: A case study from the Lower Congo Basin. *Carnets Geol.*, *16*(02), 17, 3215.

42. Hollebrands, K. F., (2003). High school students' understandings of geometric transformations in the context of a technological environment. *The Journal of Mathematical Behavior*, *22*(1), 55–72.

43. Hughes, T. J., Cottrell, J. A., & Bazilevs, Y., (2005). Isogeometric analysis: CAD, finite elements, NURBS, exact geometry and mesh

refinement. *Computer Methods in Applied Mechanics and Engineering, 194*(39–41), 4135–4195.

44. Hwang, W. Y., Chen, N. S., Dung, J. J., & Yang, Y. L., (2007). Multiple representation skills and creativity effects on mathematical problem solving using a multimedia whiteboard system. *Journal of Educational Technology & Society, 10*(2), 191–212.
45. Ismail, K., Sayed, T., & Saunier, N., (2013). A methodology for precise camera calibration for data collection applications in urban traffic scenes. *Canadian Journal of Civil Engineering, 40*(1), 57–67.
46. Iyer, N., Jayanti, S., Lou, K., Kalyanaraman, Y., & Ramani, K. (2005). Three-dimensional shape searching: state-of-the-art review and future trends. *Computer-Aided Design, 37*(5), 509-525.
47. Jacobson, R. D., & Kitchin, R. M., (1997). GIS and people with visual impairments or blindness: Exploring the potential for education, orientation, and navigation. *Transactions in GIS, 2*(4), 315–332.
48. Jafri, R., Aljuhani, A. M., & Ali, S. A., (2017). A tangible user interface-based application utilizing 3D-printed manipulatives for teaching tactual shape perception and spatial awareness sub-concepts to visually impaired children. *International Journal of Child-Computer Interaction, 11*, 3–11.
49. Jaswon, M. A., & Dove, D. B., (1960). The crystallography of deformation twinning. *Acta Crystallographica, 13*(3), 232–240.
50. Jia, Q., Fan, X., Luo, Z., Li, H., Huyan, K., & Li, Z., (2016). Cross-view action matching using a novel projective invariant on non-coplanar space-time points. *Multimedia Tools and Applications, 75*, 11661–11682.
51. Kaufmann, H., Steinbügl, K., Dünser, A., & Glück, J., (2005). General training of spatial abilities by geometry education in augmented reality. *Annual Review of Cyber Therapy and Telemedicine: A Decade of VR, 3*, 65–76.
52. Kim, H., & Lee, S., (2012). Simultaneous line matching and epipolar geometry estimation based on the intersection context of coplanar line pairs. *Pattern Recognition Letters, 33*(10), 1349–1363.
53. Klein, L. K., Maiello, G., Paulun, V. C., & Fleming, R. W., (2020). Predicting precision grip grasp locations on three-dimensional objects. *PLoS Computational Biology, 16*(8), 1008081.

54. Krumhansl, C. L., (2005). The geometry of musical structure: A brief introduction and history. *Computers in Entertainment (CIE)*, *3*(4), 1–14.

55. Kvasz, L., (1998). History of geometry and the development of the form of its language. *Synthase*, *116*, 141–186.

56. Maarif, S., Perbowo, K. S., Noto, M. S., & Harisman, Y., (2019). Obstacles in constructing geometrical proofs of mathematics-teacher-students based on Boero's proving model. In: *Journal of Physics: Conference Series* (Vol. 1315, pp. 12043). IOP Publishing.

57. Marques, E. F., Regev, O., Khan, A., & Lindman, B., (2003). Self-organization of double-chained and pseudodouble-chained surfactants: Counterion and geometry effects. *Advances in Colloid and Interface Science*, *100*, 83–104.

58. Mathai, A. M., (1999). *An Introduction to Geometrical Probability: Distributional Aspects with Applications* (Vol. 1, pp. 214–216). CRC Press.

59. Menger, K., (1971). The geometry relevant to modern education. In: *The Teaching of Geometry at the Pre-College Level: Proceedings of the Second CSMP International Conference Co-Sponsored by Southern Illinois University and Central Midwestern Regional Educational Laboratory* (Vol. 2, pp. 225–241). Dordrecht: Springer Netherlands.

60. Minsky, M., & Papert, S., (1969). An introduction to computational geometry. *Cambridge Tiass., HIT*, *479*(480), 104.

61. Moses-Snipes, P. R., (2005). The effect of African culture on African American students' achievement on selected geometry topics in the elementary mathematics classroom. *Negro Educational Review*, *56*(2/3), 147.

62. Mulligan, J., (2015). Looking within and beyond the geometry curriculum: Connecting spatial reasoning to mathematics learning. *Zdm*, *47*, 511–517.

63. Mykytiuk, S., & Shenitzer, A., (2002). Four significant axiomatic systems and some of the issues associated with them. *Mathematical Evolutions*, *219*, 6510–6516.

64. Nyerges, T. L., (1990). Locational referencing and highway segmentation in a geographic information system. *ITE Journal*, *60*(3), 27–31.

65. Odintsov, S. D., & Oikonomou, V. K., (2020). Geometric inflation and dark energy with axion F (R) gravity. *Physical Review D, 101*(4), 044009.

66. Oke, T. R., (1981). Canyon geometry and the nocturnal urban heat island: Comparison of scale model and field observations. *Journal of Climatology*, *1*(3), 237–254.

67. Ottley, A., Peck, E. M., Harrison, L. T., Afergan, D., Ziemkiewicz, C., Taylor, H. A., & Chang, R., (2015). Improving Bayesian reasoning: The effects of phrasing, visualization, and spatial ability. *IEEE Transactions on Visualization and Computer Graphics*, *22*(1), 529–538.

68. Paas, F. G., & Van Merriënboer, J. J., (1994). Variability of worked examples and transfer of geometrical problem-solving skills: A cognitive-load approach. *Journal of Educational Psychology*, *86*(1), 122.

69. Pamungkas, M. D., Rahmawati, F., & Apriliyani, M. N., (2021). Teaching spatial geometry with geogebra: Can it improve the problem-solving skills of prospective mathematics teachers? In: *Journal of Physics: Conference Series* (Vol. 1918, pp. 42082). IOP Publishing.

70. Pomeroy, E., Mushrif-Tripathy, V., Cole, T. J., Wells, J. C., & Stock, J. T., (2019). Ancient origins of low lean mass among South Asians and implications for modern type 2 diabetes susceptibility. *Scientific Reports*, *9*(1), 10515.

71. Ponce, J., Sturmfels, B., & Trager, M., (2017). Congruences and concurrent lines in multi-view geometry. *Advances in Applied Mathematics*, *88*, 62–91.

72. Pottmann, H., Peternell, M., & Ravani, B., (1999). An introduction to line geometry with applications. *Computer-Aided Design*, *31*(1), 3–16.

73. Reiss, K. M., Heinze, A., Renkl, A., & Groß, C., (2008). Reasoning and proof in geometry: Effects of a learning environment based on heuristic worked-out examples. *ZDM*, *40*(3), 455–467.

74. Rho, J., Ye, Z., Xiong, Y., Yin, X., Liu, Z., Choi, H., & Zhang, X., (2010). Spherical hyperlens for two-dimensional sub-diffractional imaging at visible frequencies. *Nature Communications*, *1*(1), 143.

75. Rian, I. M., & Asayama, S., (2016). Computational design of a nature-inspired architectural structure using the concepts of self-similar and random fractals. *Automation in Construction*, *66*, 43–58.

76. Rosen, S. M., (2017). Quantum gravity and Taoist cosmology: Exploring the ancient origins of phenomenological string theory. *Progress in Biophysics and Molecular Biology*, *131*, 34–60.

77. Russ, R. S., Scherr, R. E., Hammer, D., & Mikeska, J., (2008). Recognizing mechanistic reasoning in student scientific inquiry: A framework for discourse analysis developed from philosophy of science. *Science Education*, *92*(3), 499–525.

78. Sagawa, M., Makimoto, M., & Yamashita, S., (1997). Geometrical structures and fundamental characteristics of microwave stepped-impedance resonators. *IEEE Transactions on Microwave Theory and Techniques*, *45*(7), 1078–1085.

79. Salzman, M. C., Dede, C., Loftin, R. B., & Chen, J., (1999). A model for understanding how virtual reality aids complex conceptual learning. *Presence: Teleoperators & Virtual Environments*, *8*(3), 293–316.

80. Schardl, C. L., Craven, K. D., Speakman, S., Stromberg, A., Lindstrom, A., & Yoshida, R., (2008). A novel test for host-symbiont codivergence indicates ancient origin of fungal endophytes in grasses. *Systematic Biology*, *57*(3), 483–498.

81. Scheiter, K., Gerjets, P., & Schuh, J., (2010). The acquisition of problem-solving skills in mathematics: How animations can aid understanding of structural problem features and solution procedures. *Instructional Science*, *38*, 487–502.

82. Iyer, N., Jayanti, S., Lou, K., Kalyanaraman, Y., & Ramani, K. (2005). Three-dimensional shape searching: state-of-the-art review and future trends. *Computer-Aided Design*, *37*(5), 509-530.

83. Seidenberg, A., (1961). The ritual origin of geometry. *Archive for History of Exact Sciences*, *1*, 488–527.

84. Shu, F., Wang, J., Pagani, A., & Stricker, D., (2023). Structure PLP-slam: Efficient sparse mapping and localization using point, line, and plane for monocular, RGB-D, and stereo cameras. In: *2023 IEEE International Conference on Robotics and Automation (ICRA)* (Vol. 2, pp. 2105–2112). IEEE.

85. Siepmann, J., & Siepmann, F., (2008). Mathematical modeling of drug delivery. *International Journal of Pharmaceutics*, *364*(2), 328–343.

86. Siyepu, S. W., & Mtonjeni, T., (2014). Geometrical concepts in real-life context: A case of South African traffic road signs. In: *Proceedings of the 20th Annual National Congress of the Association for Mathematics Education of South Africa* (Vol. 1, pp. 41–45).

87. Snijders, C. J., Vleeming, A., & Stoeckart, R., (1993). Transfer of lumbosacral load to iliac bones and legs: Part 1: Biomechanics of self-bracing of the sacroiliac joints and its significance for treatment and exercise. *Clinical Biomechanics*, *8*(6), 285–294.
88. Stieff, M., (2011). When is a molecule three dimensional? A task-specific role for imagistic reasoning in advanced chemistry. *Science Education*, *95*(2), 310–336.
89. Sullivan, R., (2001). Deformity: A modern western prejudice with ancient origins. *Proceedings-Royal College of Physicians of Edinburgh*, *31*(3), 262–266.
90. Sunzuma, G., & Maharaj, A., (2022). Zimbabwean in-service teachers' views of geometry: An ethnomathematics perspective. *International Journal of Mathematical Education in Science and Technology*, *53*(9), 2504–2515.
91. Tang, G., (2015). An overview of historical and contemporary concrete shells, their construction and factors in their general disappearance. *International Journal of Space Structures*, *30*(1), 1–12.
92. Treeby, B. E., & Cox, B. T., (2010). k-Wave: MATLAB toolbox for the simulation and reconstruction of photoacoustic wave fields. *Journal of Biomedical Optics*, *15*(2), 021314–021314.
93. Trofimov, V. V., (2013). *Introduction to Geometry of Manifolds with Symmetry* (Vol. 270, pp. 1–05). Springer Science & Business Media.
94. Turrin, M., Von Buelow, P., & Stouffs, R., (2011). Design explorations of performance driven geometry in architectural design using parametric modeling and genetic algorithms. *Advanced Engineering Informatics*, *25*(4), 656–675.
95. Vavra, G., (1990). On the kinematics of zircon growth and its petrogenetic significance: A cathodoluminescence study. *Contributions to Mineralogy and Petrology*, *106*(1), 90–99.
96. Verdine, B. N., Lucca, K. R., Golinkoff, R. M., Hirsh-Pasek, K., & Newcombe, N. S., (2016). The shape of things: The origin of young children's knowledge of the names and properties of geometric forms. *Journal of Cognition and Development*, *17*(1), 142–161.
97. Wagner, L., Box, H., & Morehead, S. K., (2013). *Ancient Origins of the Mexican Plaza: From Primordial Sea to Public Space* (Vol. 54, No. 4, pp. 6540–6545). University of Texas Press.

98. Wang, Y., Zhang, H., Zhang, X., & Wang, X., (2022). Method for extrinsic parameters calibration of Pan-Tilt camera using intersecting lines from two images. *Measurement*, *202*, 111791.
99. Wylie, R. G., (1952). The properties of liquid phase embryos at the intersections of plane solid surfaces. *Australian Journal of Chemistry*, *5*(4), 618–627.
100. Yang, S. Q., Tian, W. L., Huang, Y. H., Ranjith, P. G., & Ju, Y., (2016). An experimental and numerical study on cracking behavior of brittle sandstone containing two non-coplanar fissures under uniaxial compression. *Rock Mechanics and Rock Engineering*, *49*, 1497–1515.
101. Yao, X., & Manouchehri, A., (2019). Middle school students' generalizations about properties of geometric transformations in a dynamic geometry environment. *The Journal of Mathematical Behavior*, *55*, 100703.
102. Yariv, L., Kasten, Y., Moran, D., Galun, M., Atzmon, M., Ronen, B., & Lipman, Y., (2020). Multiview neural surface reconstruction by disentangling geometry and appearance. *Advances in Neural Information Processing Systems*, *33*, 2492–2502.
103. Ye, T., Mittal, R., Udaykumar, H. S., & Shyy, W., (1999). An accurate Cartesian grid method for viscous incompressible flows with complex immersed boundaries. *Journal of Computational Physics*, *156*(2), 209–240.
104. Zakharov, P. V., Vdovin, R. S., Markidonov, A. V., Kochkin, A. S., & Vdovin, A. S., (2020). Virtual and mixed reality in the study of the geometry of the crystal lattice. In: *Journal of Physics: Conference Series* (Vol. 1515, pp. 2200–2204). IOP Publishing.

CHAPTER 2

Fundamentals of Shapes

Contents

2.1. INTRODUCTION

Geometric shapes are fundamental elements in geometry, building blocks for understanding the three-dimensional world surrounding us. This chapter will explore the basics of geometric shapes by exploring their definitions and discussing their significance.

Geometric shapes, also called geometric figures or simply shapes, are identifiable, well-defined objects with distinct characteristics and properties. These shapes are typically two-dimensional (2D) or three-dimensional (3D) entities that can be precisely described using mathematical language and equations. 2D geometric shapes include circles, triangles, rectangles, and polygons, while 3D shapes encompass spheres, cubes, cylinders, and pyramids (Wu et al., 2015).

Mathematically, a geometric shape can be defined as a set of points in space that satisfies certain conditions or constraints. These conditions often involve relationships between the shape's components, such as its sides, angles, and dimensions. For instance, a circle can be defined as the set of all points equidistant from a central point, known as the center, while its three sides and three angles define a triangle.

It is essential to emphasize that geometric shapes, as mathematical abstractions, are idealized representations of real-world objects. They serve as simplified models that allow us to analyze and understand the underlying principles governing various physical phenomena. The geometric shapes we encounter in geometry are devoid of imperfections, possessing exact measurements and idealized characteristics, making them valuable tools for mathematical analysis (Huang et al., 2012).

2.1.1. Importance of Understanding Shapes

Geometrical shapes play a pivotal role in mathematics and science, providing a framework for understanding spatial concepts, solving practical problems, and exploring the intricacies of our physical world. Understanding geometric shapes is fundamental to geometry for several reasons (Elia & Gagatsis, 2003):

1. **Foundation of Spatial Reasoning:** Geometric shapes are the foundation for developing spatial reasoning skills. They enable us to visualize, analyze, and manipulate objects in space, fostering critical thinking and problem-solving abilities.
2. **Geometry's Practical Applications:** Geometry has countless practical applications in diverse fields, such as architecture,

engineering, art, and design. A solid grasp of geometric shapes is essential for professionals in these disciplines to create accurate plans, designs, and structures.

3. **Mathematical Principles:** Geometric shapes embody fundamental mathematical principles, including symmetry, congruence, similarity, and transformation. These principles are essential for understanding complex geometric concepts and proofs.
4. **Real-World Problem Solving:** Many real-world problems involve spatial relationships and measurements. Geometry equips us with the tools to address these problems, from calculating objects' dimensions to designing efficient navigation routes.

2.2. CLASSIFICATION OF GEOMETRIC SHAPES

Classification of geometric shapes is an essential aspect of geometry that categorizes them based on their defining characteristics and properties. This categorization simplifies the study of shapes by grouping them into distinct classes, making it easier to analyze their common features and differences. Geometric shapes can be classified in various ways, including by the number of sides, angles, and dimensions they possess. This systematic approach to classification aids in understanding the relationships and patterns among different shapes, paving the way for a deeper exploration of geometry's intricacies (Kirner et al., 2012).

2.2.1. Two-Dimensional Shapes

In geometry, two-dimensional (2D) shapes are flat figures that exist only in two dimensions: length and width. These shapes are crucial in understanding spatial relationships and are commonly encountered in real-world applications. Here, we explore some fundamental 2D shapes: circles, triangles, quadrilaterals, and polygons (Kimia et al., 1995).

Tip

When exploring geometric transformations, remember that reflections are like flipping a shape over a line, rotations involve turning a shape around a fixed point, and translations mean shifting a shape while maintaining its size and shape.

2.2.1.1. Circles

A circle is a perfectly round shape with all points on its boundary equidistant from its center. The distance from the center to any point on the circle is called the radius (r), and the longest chord that passes through the center is the diameter (d), which is twice the radius (d = 2r). A circle's circumference (C) can be calculated using the formula (Anell & Midtkandal, 2017).

$$C = 2pr$$

where; π (pi) is approximately equal to 3.14159. The area (A) of a circle can be calculated using the formula:

$$A = pr^2$$

Circles are essential in geometry, trigonometry, and calculus. They have numerous real-world applications, from calculating the area of circular objects to understanding the orbits of celestial bodies (Figure 2.1).

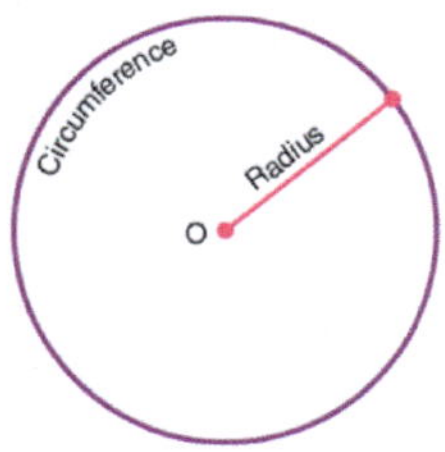

Figure 2.1. Schematic of a 2D circle.

Source: BYJU's, Creative Commons License.

2.2.1.2. Triangles

A triangle is a polygon with three sides, angles, and vertices. Triangles are versatile geometric shapes and come in various types based on their sides and angles. Some common types include (Borissova et al., 2022):

i. **Equilateral Triangle:** All sides and angles are equal.
ii. **Isosceles Triangle:** Two sides and two angles are equal.
iii. **Scalene Triangle:** No sides or angles are equal.
iv. **Acute Triangle:** All angles are less than 90°.
v. **Obtuse Triangle:** One angle is greater than 90°.
vi. **Right Triangle:** One angle is equal to 90° (Figure 2.2).

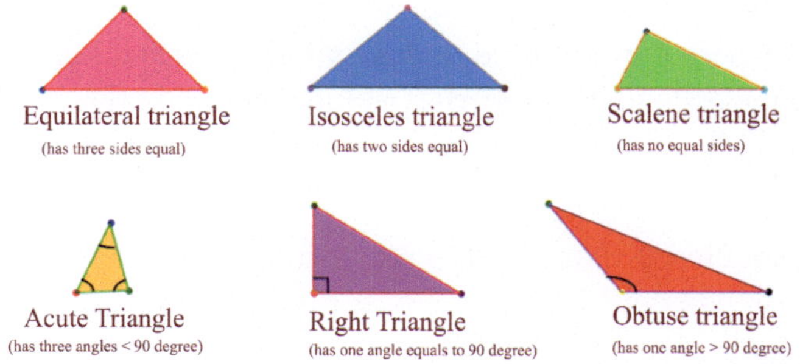

Figure 2.2. Illustration of various types of triangles.

Source: Play Worksheet, Creative Commons License.

The sum of the interior angles of any triangle is always 180°. Triangles find applications in trigonometry, engineering, and architecture, as they serve as the basis for understanding complex structures and angles.

2.2.1.3. Quadrilaterals

Quadrilaterals are polygons with four sides, four angles, and four vertices. They come in various forms, each with its unique properties. Some common types of quadrilaterals include (Drake, 2011):

i. **Rectangle:** All angles are right angles (90°).
ii. **Square:** All sides are equal, and all angles are right angles.
iii. **Parallelogram:** Opposite sides are parallel, and opposite angles are equal.
iv. **Rhombus:** All sides are equal, and opposite angles are equal.
v. **Trapezoid:** One pair of opposite sides is parallel.
vi. **Kite:** Two pairs of adjacent sides are equal (Figure 2.3).

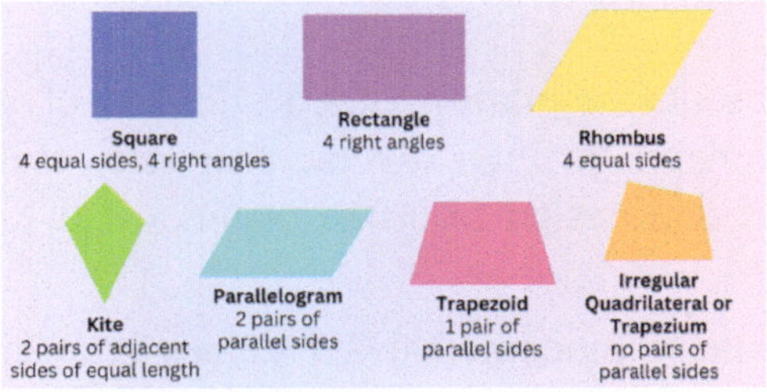

Figure 2.3. Different shapes of quadrilaterals.

Source: Science Notes, Creative Commons License.

Quadrilaterals have wide-ranging applications in fields such as architecture, where they are used to design buildings and structures, and in geometry to understand the properties of more complex shapes.

2.2.1.4. Polygons

A polygon is a closed figure with three or more straight sides and angles. Some common polygons include (Low & List, 1982) (Figure 2.4):

i. **Pentagon:** A five-sided polygon.
ii. **Hexagon:** A six-sided polygon.
iii. **Heptagon:** A seven-sided polygon.
iv. **Octagon:** An eight-sided polygon.
v. **Decagon:** A 10-sided polygon.

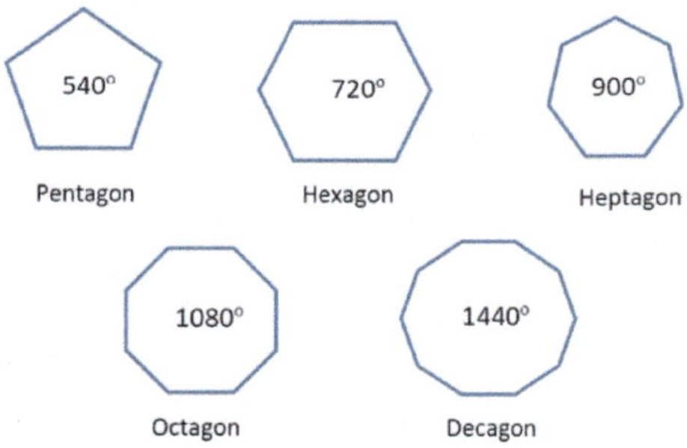

Figure 2.4. Illustration of different polygons.

Source: Lifewire, Creative Commons License.

The sum of the interior angles of a polygon can be calculated using the formula:

$$Sum\ of\ interior\ angles = (n-2)\times 180^{\circ}$$

Here; "n" represents the number of sides in the polygon. Polygons are widely used in geometry for their diverse properties and can be found in everything from art and design to urban planning (El Korchi & Ghanou, 2020).

Understanding these fundamental 2D shapes is essential for grasping more complex geometric concepts and solving practical problems in various fields, from mathematics and science to architecture and engineering.

2.2.2. Three-Dimensional Shapes

In geometry, three-dimensional (3D) shapes are solid figures in three dimensions: length, width, and height. These shapes are crucial in understanding spatial relationships and are prevalent in various real-world applications. Here, we explore some fundamental 3D shapes: spheres, cubes, cylinders, and pyramids (Rege et al., 2013) (Figure 2.5).

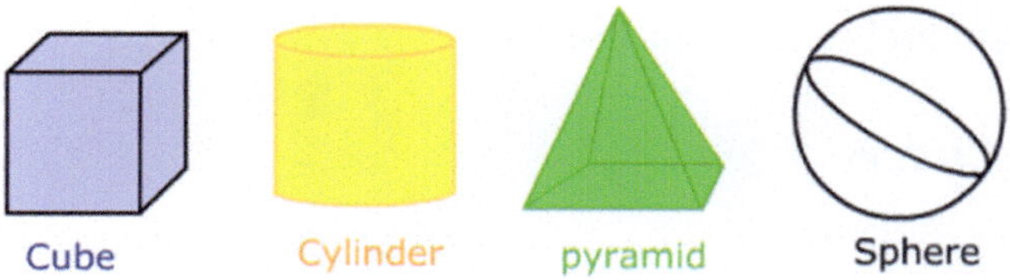

Figure 2.5. Schematic of various 3D shapes.

Source: Soft Schools, Creative Commons License.

2.2.2.1. Spheres

A sphere is a perfectly round 3D shape that resembles a ball. It has all points on its surface equidistant from its center. The distance from the center to any point on the sphere is called the radius (r). Spheres are known for their unique properties. The volume (V) of a sphere can be calculated using the formula:

$$V = \frac{4}{3} p r^3$$

where; π (pi) is approximately equal to 3.14159. The surface area (A) of a sphere can be calculated using the formula:

$$A = 4 p r^2$$

Spheres are prevalent in science and engineering, such as in modeling celestial bodies like planets and stars or in the design of spherical structures.

2.2.2.2. Cubes

A cube is a 3D shape with six equal square faces, 12 edges, and eight vertices. All angles in a cube are right angles (90°), and all sides are equal in length. The volume (V) of a cube can be calculated using the formula:

$$V = a^3$$

where; "a" represents the length of one side of the cube. The surface area (A) of a cube can be calculated using the formula:

$$A = 6a^2$$

Cubes are fundamental in geometry and mathematics, often used to introduce concepts like volume and surface area. They also have practical applications in architecture and design.

2.2.2.3. Cylinders

A cylinder is a 3D shape with two circular bases connected by a curved surface. The volume (V) of a cylinder can be calculated using the formula (Tangelder & Veltkamp, 2008):

$$V = \rho r^2 h$$

where; "r" represents the radius of the base; and "h" represents the height. The lateral surface area (LSA) of a cylinder can be calculated using the formula:

$$LSA = 2\rho rh$$

Cylinders are commonly found in everyday objects like cans, pipes, and containers. Understanding their volume and surface area is essential for various engineering and design applications.

2.2.2.4. Pyramids

A pyramid is a 3D shape with a polygonal base and triangular faces that meet at a common vertex called the apex. The volume (V) of a pyramid can be calculated using the formula:

$$V = \frac{1}{3} Bh$$

where; "B" represents the base area; and "h" represents the height from the apex to the base. The surface area (A) of a pyramid can vary depending on the type of pyramid, but it typically includes the area of the base and the areas of the triangular faces (Hyde, 2002).

Pyramids are significant in geometry and architecture, as they appear in the design of ancient structures like the pyramids of Egypt and modern buildings.

2.2.3. Irregular Shapes

While basic geometric shapes like circles, triangles, and cubes have well-defined properties, irregular shapes are more complex and lack regularity in their form. In this section, we will study the definition of irregular shapes and explore various examples to understand their characteristics and significance in geometry (Pizlo et al., 2010).

Irregular or complex shapes do not conform to standard geometric definitions. They lack symmetry, uniformity, or regular angles in basic shapes. Irregular shapes can have sides of varying lengths and angles, often exhibiting a more intricate structure. These shapes challenge our understanding of geometry due to their diversity and unique attributes.

2.2.3.1. Examples of Irregular Shapes

Following are the examples of the irregular shapes (Brechbühler et al., 1995):

1. **Irregular Polygons:** One of the most common types of irregular shapes is the irregular polygon. Unlike regular polygons (e.g., squares or equilateral triangles), irregular polygons have sides of different lengths and angles that may not be congruent. For instance, a pentagon with sides of varying lengths and non-right angles is an irregular polygon.
2. **Irregular Quadrilaterals:** Quadrilaterals are shapes with four sides. While rectangles and squares are regular quadrilaterals, irregular quadrilaterals encompass many shapes. A trapezoid with only one pair of parallel sides and unequal side lengths is an example of an irregular quadrilateral.
3. **Irregular Curves:** Some irregular shapes are curved and do not follow the straight lines typically associated with basic shapes. Examples include the shape of a coastline, a meandering river, or the outline of a cloud.
4. **Complex Figures:** Irregular shapes can also be formed by combining basic shapes or irregular polygons. For instance, a figure composed of a rectangle and a triangle attached is irregular. Similarly, irregular shapes can arise in composite structures like architectural designs or irregularly shaped objects in real-life scenarios (Figure 2.6).

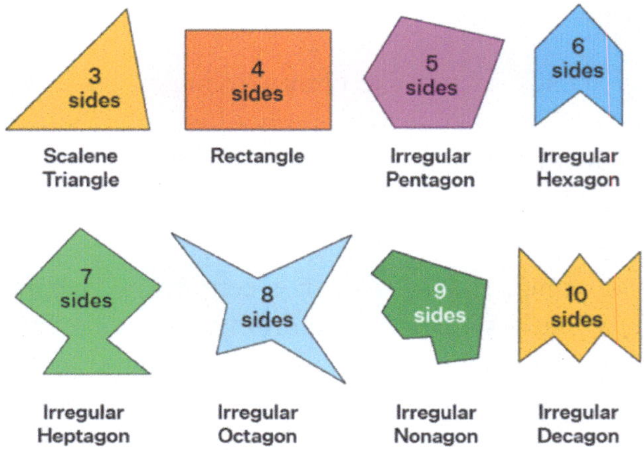

Figure 2.6. Illustration of various irregular polygons.

Source: Cuemath, Creative Commons License.

2.2.3.2. Characteristics of Irregular Shapes

The following are the characteristics of irregular shapes (İbili et al., 2020):

1. **Lack of Symmetry:** Irregular shapes often lack the symmetry commonly found in basic shapes. They may not have a line of symmetry or rotational symmetry.
2. **Variability:** Irregular shapes exhibit significant side lengths, angles, and proportions variability. This variability makes it challenging to classify or analyze shapes using traditional geometric methods.
3. **Unique Properties:** Each irregular shape has unique properties and may require specific approaches for measurement and analysis. There are no universal formulas for the properties of irregular shapes as there are for regular ones.
4. **Real-World Relevance:** Irregular shapes are prevalent in the real world. Natural features like coastlines, clouds, and mountain ranges often have irregular shapes. Manufactured objects, architectural designs, and geographic boundaries also frequently exhibit irregularity.

Hence, these shapes require flexible problem-solving skills and mathematical techniques to accurately analyze and describe their properties.

Whether in cartography, art, architecture, or engineering, irregular shapes play a significant role in our understanding of spatial relationships and design.

2.3. PROPERTIES OF GEOMETRIC SHAPES

Geometric shapes possess distinctive properties, including attributes like symmetry, angles, sides, and dimensions. For instance, triangles can be classified based on their angles as acute, obtuse, or right, while rectangles exhibit perpendicular angles and equal opposite sides. Understanding these properties is fundamental for analyzing and manipulating shapes in various mathematical and practical contexts. These properties aid in classification and serve as the foundation for solving geometric problems, whether calculating areas, finding angles, or exploring spatial relationships (Zerroug & Nevatia, 1993).

Remember

Symmetry plays a crucial role in understanding shapes. Line symmetry refers to shapes that can be folded along a line, with both sides matching. Rotational symmetry means a shape can be rotated to match itself at certain angles.

When delving into the properties of geometric shapes, it is crucial to focus on some fundamental forms, such as circles, triangles, and rectangles. Each of these shapes exhibits distinct characteristics that make them essential elements in geometry:

1. **Circles:** Their primary characteristic is their radial symmetry, meaning that any line drawn from the center to the edge (the radius) is the same length. Circles are characterized by their radius (r) and diameter (d), with the relationship:

$$d = 2r.$$

Their area (A) is calculated using the formula:

$$A = pr^2$$

While their circumference (C) is given by:

$$C = 2pr$$

2. **Triangles:** These have unique properties like the Pythagorean theorem, which relates the lengths of their sides:

$$a^2 + b^2 = c^2$$

where; 'a' and 'b' are the shorter sides (legs) and 'c' is the longest (hypotenuse).

3. **Rectangles:** These are quadrilaterals with four right angles (90°). They possess pairs of opposite sides that are equal in length, and their diagonals bisect each other at right angles. The area (A) of a rectangle can be calculated using the formula:

$$A = length \times width$$

Its perimeter (P) is given by:

$$P = 2(length + width)$$

Hence, these properties not only aid in shape identification but also play a pivotal role in solving geometric problems and real-world applications, from calculating the area of a circular garden to determining the angles of a triangular roof truss.

2.3.1. Understanding Dimensions

In geometry, understanding dimensions is essential when dealing with various geometric shapes. The dimensions of a shape refer to its measurable attributes that define its size and extent in space. Key dimensions include length, width, height, radius, and diameter, each playing a distinct role in different geometric contexts (Dogruer & Akyuz, 2020):

1. **Length:** It is a linear dimension that measures the extent of an object or shape in one direction. In geometric terms, it is typically associated with the longest side of a shape. For example, in a rectangle, the length is the longer of the two sides, and in a line segment, it represents the entire span between two points.
2. **Width:** Also, a linear dimension, measures the extent of an object or shape perpendicular to its length. In rectangles, the width is the shorter side. For example, in a standard piece of paper, the length is longer (typically 11 inches), and the width is shorter (usually 8.5 inches).
3. **Height:** It is the vertical dimension that measures the distance from the base of an object or shape to its highest point. In geometric terms, height is commonly associated with three-dimensional shapes like cylinders or prisms. For instance, the height of a rectangular prism is the vertical distance between its top and bottom faces.

4. **Radius:** It is a measure associated primarily with circular shapes, like circles and spheres. It represents the distance from the center of a circle to any point on its circumference. In equations for circles, the radius is often denoted as 'r.' For example, if you have a circle with a radius of 5 units, its equation would be:

$$x^2 + y^2 = 5^2.$$

5. **Diameter:** The diameter of a circle is a linear dimension that measures the distance across the circle, passing through its center. It is precisely twice the length of the radius. In equations, the diameter is represented as 'd,' and it's related to the radius by the formula d = 2r.

Thus, understanding dimensions is vital for various geometric calculations. For example, when finding the volume of a cylindrical container, you need to know both the radius and height. Similarly, in determining the area of a rectangular plot of land, you would require length and width. These dimensions provide the numerical values necessary for precise geometric calculations and real-world applications, ensuring that geometry remains a practical and invaluable branch of mathematics.

2.3.2. Properties of Polygons

Polygons are a fundamental class of geometric shapes with distinct properties related to their angles, sides, and perimeter. Understanding these properties is crucial in geometry, enabling precise measurements and calculations (Zhong et al., 2019).

1. **Angles of Polygons:** The angles in a polygon play a significant role in its classification and properties. The formula can determine the sum of interior angles in any polygon:

$$Sum\ of\ interior\ angles = (n-2)\cdot 180^{\circ}$$

Here, 'n' represents the number of sides or vertices in the polygon. For example, a triangle (a polygon with three sides) has a sum of interior angles equal to 180°, while a hexagon (six sides) has a sum of 720°.

Furthermore, regular polygons, where all sides and angles are congruent, have a unique property. In a regular polygon with 'n' sides, each interior angle measures:

$$Interior\ angle\,(Regular\ Polygon) = \frac{360^{\circ}}{n}$$

For instance, in a regular pentagon, each interior angle measures 72°.

2. **Sides of Polygons:** These are line segments connecting consecutive vertices. The number of sides determines the polygon's classification. Some common polygons include triangles (3 sides), quadrilaterals (4 sides), pentagons (5 sides), and hexagons (6 sides). Polygons with more sides are referred to as polygons with 'n' sides, such as octagons (8 sides) and decagons (10 sides) (Ngom et al., 2012).
3. **Perimeter of Polygons:** The perimeter of a polygon is the sum of the lengths of its sides. The formula to calculate the perimeter depends on the type of polygon.

For a regular polygon with 'n' sides, where each side has a length 's,' the perimeter is given by:

$$Perimeter\left(Regular\ Polygon\right) = n \cdot s$$

For irregular polygons, where side lengths may vary, you sum the lengths of all sides individually.

Calculating the perimeter is essential in practical applications. For example, determining the fencing required to enclose a polygonal garden or finding the distance traveled along a polygonal path.

2.3.3. Volume and Surface Area for 3D Shapes

We will explore the formulas and calculations for these two critical properties (Pedai et al., 2021).

2.3.3.1. The Volume of 3D Shapes

1. **Cuboids and Rectangular Prisms:** These are among the simplest 3D shapes. The volume $\left(V\right)$ A cuboid or rectangular prism can be calculated using the formula:

$$V = l \cdot w \cdot h$$

where; l represents the length; w is the width; and h stands for the height of the shape.

2. **Spheres:** The volume (V) of a sphere with a radius $\left(r\right)$ is given by:

$$V = \frac{4}{3}ðr^3$$

3. **Cylinders:** For a cylinder with a base radius (r) and height (h), the volume (V) is calculated as:

$$V = ðr^2h$$

2.3.4. Surface Area of 3D Shapes

1. **Cuboids and Rectangular Prisms:** The surface area (SA)*of these* shapes involve adding the areas of all six faces. The formula is:

$$SA = 2lw + 2lh + 2wh$$

Each term represents the area of one pair of opposite faces.

2. **Spheres:** The surface area (SA) of a sphere is given by:

$$SA = 4\pi r^2$$

3. **Cylinders:** The surface area (SA) of a cylinder can be calculated by adding the areas of its two circular bases and the lateral surface area. The formula is:

$$SA = 2\pi r^2 + 2\pi rh$$

here; $(2\pi rh)$ represents the areas of the bases; and $(2\pi r^2)$ is the lateral surface area.

These volume and surface area formulas are foundational for solving geometric problems involving 3D shapes. For instance, in construction, understanding the volume helps determine the amount of material needed to build a structure, while calculating surface area aids in estimating the quantity of paint or other surface coatings required (Hyde, 2002).

2.4. CONGRUENCE AND SIMILARITY

Congruence and similarity are vital for geometric proofs, transformations, and real-world applications, such as map scaling and architectural design. Understanding congruence and similarity allows mathematicians and scientists to analyze and work with shapes, making these concepts crucial in geometry (Tangelder & Veltkamp, 2008) (Figure 2.7).

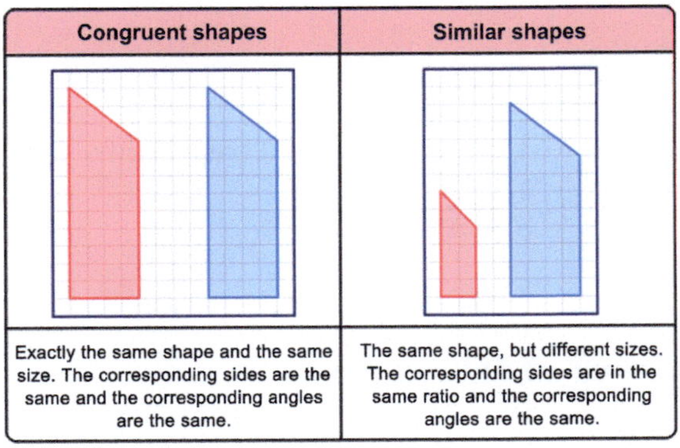

Figure 2.7. Difference between congruence and similarity in geometry.

Source: Third Space Learning, Creative Commons License.

Did You Know?

Many everyday objects can be analyzed and understood using geometric formulas. For instance, the formula for the area of a circle, $A = \pi r^2$, is essential in various fields, including architecture and engineering.

2.4.1. Congruent Shapes

Certain criteria must be met to establish congruence between shapes, and congruence transformations are essential for demonstrating congruence.

2.4.1.1. Definition and Criteria

Two shapes are congruent if all their corresponding angles and sides are equal. It means they match when you superimpose one shape onto the other. The criteria for congruence can be summarized through the following (Kirner et al., 2012):

1. Corresponding Angles: All angles in one shape must have their corresponding angles in the other shape equal. For example, if angle A in shape X corresponds to angle A' in shape Y, then A = A.'
2. **Corresponding Sides:** All sides in one shape must have their corresponding sides in the other shape equal in length. For instance, if side AB in shape X corresponds to side A'B' in shape Y, then AB = A'B.'

3. **Congruence of Triangles:** When dealing with triangles, additional criteria include the congruence of three corresponding sides (SSS), two sides and the included angle (SAS), or two angles and the included side (ASA) (Figure 2.8).

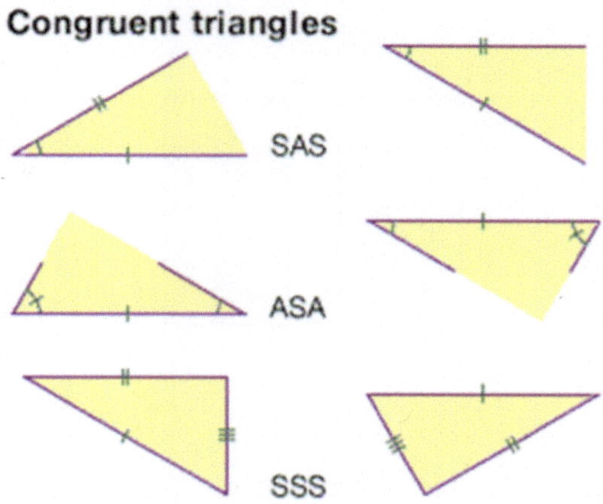

Figure 2.8. Schematic of congruent triangles.

Source: Britannica, Creative Commons License.

2.4.1.2. Congruence Transformations

To prove that two shapes are congruent, congruence transformations are applied. These transformations are specific actions that can be performed on a shape without altering its size or shape. The three primary congruence transformations are (Markov et al., 2020):

1. **Translation (T):** This involves moving a shape from one location to another, maintaining the same orientation. It preserves both angles and side lengths. For example, if you translate the shape X horizontally and vertically, it remains congruent to its original position.
2. **Rotation (R):** It transforms a shape by turning it around a fixed point (the center of rotation). It preserves angles and side lengths if the center of rotation remains the same.
3. **Reflection (F):** It flips a shape over a line, known as the line of reflection. This transformation keeps angles intact but may change side orientations. However, the side lengths remain unchanged.

These congruence transformations can be used in combination to establish congruence between shapes. For instance, if you can apply a series of rotations and reflections to shape X to superimpose it perfectly onto shape Y, you can conclude that X and Y are congruent (Figure 2.9).

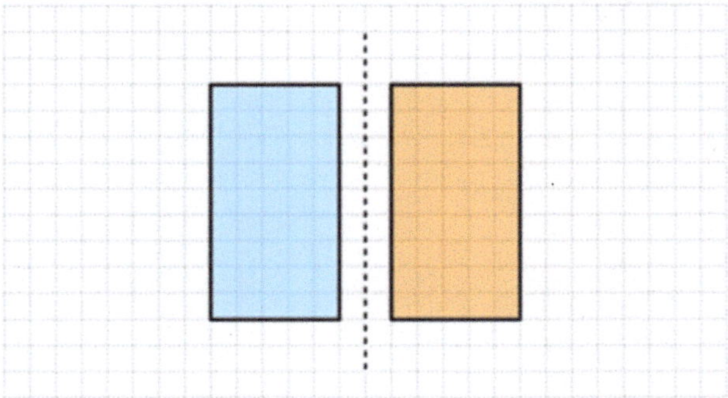

Figure 2.9. Illustration of reflected rectangles as congruent shapes.

Source: Bite-size, Creative Commons License.

2.4.2. Similar Shapes

Similarity is a concept that describes the relationship between two shapes that may not be identical in size but have the same shape. It means their angles are equal, and their corresponding sides are proportional. Understanding similar shapes and their applications in real-life situations is fundamental in various fields, from architecture to engineering (Ngom et al., 2012) (Figure 2.10).

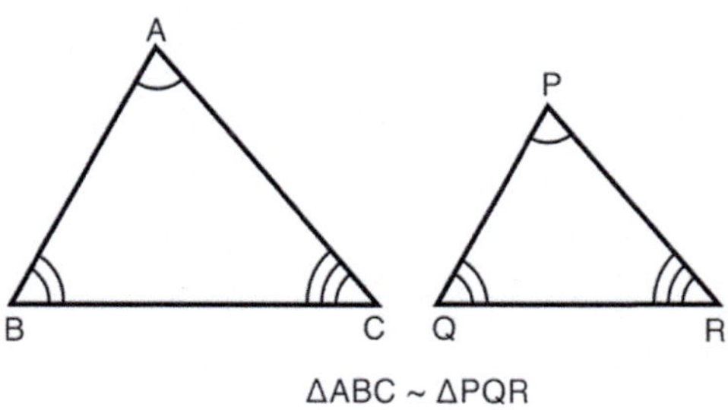

Figure 2.10. *Illustration of similar triangles* $\triangle ABC \sim \triangle PQR$.

Source: Math Monks, Creative Commons License.

2.4.2.1. Definition and Criteria

Two shapes are considered similar if they meet two main criteria (Putra et al., 2021):

1. **Corresponding Angles:** All angles in one shape are equal to their corresponding angles in the other shape. It means that if angle A in shape X corresponds to angle A' in shape Y, then A = A,' angle B in shape X corresponds to angle B' in shape Y, and so on.
2. **Proportional Sides:** The lengths of the corresponding sides of the two shapes are in proportion. If side AB in shape X corresponds to side A'B' in shape Y, then the ratio of the lengths of these sides is constant: AB/A'B' = k, where k is a constant.

It is important to note that similarity does not require the two shapes to be identical in size. Instead, it focuses on their proportional relationships.

2.4.2.2. Real-Life Example: Map Scaling

Let's consider a practical example of similarity in map scaling:

Ø **Example:** Imagine you have a map of a city, and you want to create a smaller, more portable version of the same map for tourists. To do It, you scale down the map while maintaining the proportions of streets, landmarks, and distances (Iwamoto & Ou-Yang, 2004).

1. **Select a Scale Factor:** You choose a scale factor, say 1:10,000, which means that 1 unit on the new map corresponds to 10,000 units.
2. **Maintain Angle Measures:** All the angles on the scaled-down map are the same as those on the original map. It ensures that streets still meet at intersections at the same angles.
3. **Proportional Sides:** The lengths of streets and distances between landmarks are reduced by 10,000 while maintaining proportionality. For example, if the distance between two landmarks on the original map were 5 cm, it would be 0.0005 cm (or 0.05 mm) on the scaled-down map.

By following these principles of similarity, you can create a smaller map that accurately represents the city's layout and features in proportion to the actual city. In conclusion, understanding the concept of similarity in geometry is essential for solving geometric problems and has widespread applications in various fields. It allows us to scale, compare, and work with objects and structures in a proportional and meaningful way, making it a fundamental concept in geometry with real-life significance.

2.5. GEOMETRIC TRANSFORMATIONS

Geometric transformations are fundamental operations that change geometric shapes' position, size, or orientation while preserving their essential properties. These transformations include translations, rotations, reflections, and dilations. They play a crucial role in geometry and have practical applications in computer graphics, engineering, and design. Transformations help us analyze the relationships between shapes, create symmetrical patterns, and simulate movements in the virtual world. Understanding these transformations enhances our ability to work with geometric objects and solve complex problems in various domains (Huang et al., 2012).

2.5.1. Reflections

Reflection is a fundamental geometric transformation that involves flipping a shape over a line, creating a mirror image of the original shape. This transformation preserves distances between points and angles, making it an essential concept in geometry and various applications, from art to engineering. This discussion will explore the principles behind reflections, the mathematical equations that govern them, and real-world examples to illustrate their significance.

2.5.1.1. Principles of Reflection

Reflection mirrors a shape over a specified line, called the "line of reflection" or the "mirror line."

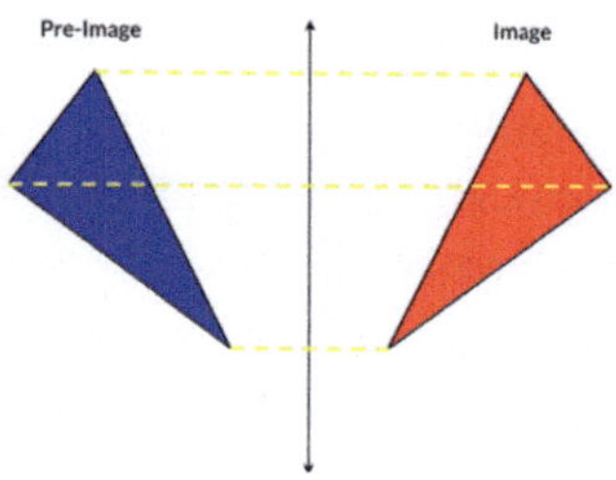

Figure 2.11. Illustration of reflection of images.

Source: Wisconsin Department of Public Instruction, Creative Commons License.

This line can be vertical, horizontal, or diagonal. When a shape is reflected, each point on the original shape is moved to an equal distance on the opposite side of the line of reflection. This distance is always perpendicular to the line of reflection, creating a perfect mirror image (Elia & Gagatsis, 2003). Mathematically, we can define the reflection of a point (x, y) over a vertical line as (–x, y) and over a horizontal line as (x, –y). We must consider the x and y components separately for a diagonal line (Figure 2.11).

2.5.1.2. Equations for Reflection

Let's consider the reflection of a point (x, y) over the x-axis and the y-axis:

1. Reflection Over the x-Axis:
   i. The x-coordinate remains the same;
   ii. The y-coordinate changes sign;
   iii. The reflection of (x, y) over the x-axis becomes (x, –y).
2. Reflection Over the y-Axis:
   i. The x-coordinate changes sign;
   ii. The y-coordinate remains the same;
   iii. The reflection of (x, y) over the y-axis becomes (–x, y).

For diagonal lines, we would need to define specific equations based on the angle and orientation of the line.

2.5.1.3. Real-World Applications

Reflections are not just theoretical concepts; they have practical applications in various fields (Drake, 2020):

1. **Architecture and Design:**
   i. Architects use reflections to create symmetrical and aesthetically pleasing designs;
   ii. Interior designers use mirrors to make spaces appear larger.
2. **Engineering:**
   i. In mechanical engineering, reflections help design symmetric components;
   ii. In civil engineering, they are used in bridge and building designs.
3. **Art and Photography:**
   i. Artists often use reflections to create balance and harmony in their compositions;

ii. Photographers use reflections in water bodies, mirrors, or glass to capture striking images.

4. **Optics:** Including lenses and mirrors, rely on reflection principles to manipulate light.

5. **Mathematics and Geometry:**

i. Reflections are fundamental in understanding geometric properties;

ii. They play a crucial role in solving problems related to symmetry and congruence.

Ø **Examples:** Let's consider a few examples to illustrate the concept of reflections:

- **Example 1: Reflection Over the x-Axis:** Suppose we have point A (3, 4). When reflected over the x-axis, it becomes A'(3, –4), creating a mirror image below the x-axis.
- **Example 2: Reflection On the y-Axis:** Let's take point B (2, –5). When reflected over the y-axis, it becomes B'(–2, –5), mirroring it across the y-axis.
- **Example 3: Artistic Reflection:** In a piece of art, if you draw a heart shape on one side of the canvas and then reflect it over a vertical line (the line of reflection), you will create a perfectly symmetrical heart.

Hence, reflections are a fundamental geometric transformation crucial in various real-world applications. Understanding the principles and mathematical equations behind reflections helps us appreciate their significance in geometry and beyond.

They allow us to create symmetry, balance, and aesthetically pleasing designs, making them an essential concept in shapes and spatial relationships.

2.5.2. Rotations: Rotating Shapes Around a Point

Rotations are fundamental geometric transformations that involve turning or spinning a shape around a fixed point called the "center of rotation."

This transformation is prevalent in geometry, art, engineering, and everyday life. This discussion will study the principles behind rotations, the mathematical equations governing them, and real-world examples demonstrating their significance (Boissonnat, 1984).

2.5.2.1. Principles of Rotations

A rotation is a transformation that takes a shape and turns it around a fixed point by a certain angle in a plane. The fixed point, or center of rotation, remains unchanged after the rotation. All other points on the shape move in a circular path around this center point. A positive angle of rotation signifies a counterclockwise turn, while a negative angle indicates a clockwise turn (Anell & Midtkandal, 2017) (Figure 2.12).

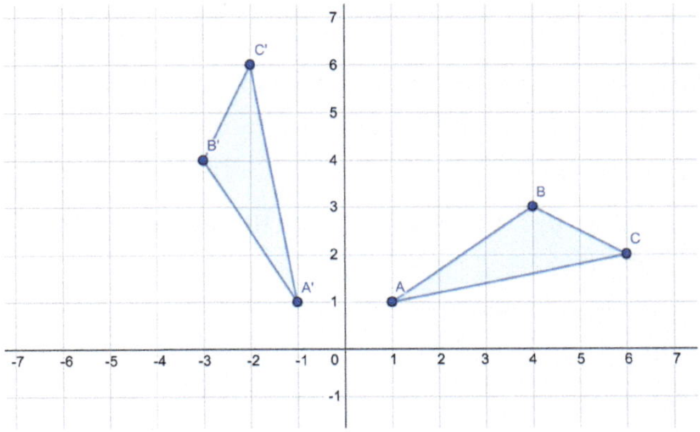

Figure 2.12. Schematic of the rotation.

Source: Story of Mathematics, Creative Commons License.

Mathematically, a rotation can be described by specifying the center of rotation (h, k) and the rotation angle θ.

2.5.2.2. Equations for Rotations

To perform a rotation around the origin (0, 0), the new coordinates (x,' y') of a point (x, y) are given by:

$$x' = x \cdot \cos(\theta) - y \cdot \sin(\theta)$$

$$y' = x \cdot \sin(\theta) + y \cdot \cos(\theta)$$

To rotate around a point (h, k), the new coordinates are:

$$x' = h + (x-h) \cdot \cos(\theta) - (y-k) \cdot \sin(\theta)$$

$$y' = k + (x-h) \cdot \sin(\theta) + (y-k) \cdot \cos(\theta)$$

Here, (x, y) are the original coordinates; (x,' y') are the rotated coordinates; (h, k) is the center of rotation; and θ is the angle of rotation.

2.5.2.3. Real-World Applications

Rotations have numerous practical applications across various domains (Alexiadis, 2006):

1. **Engineering and Mechanics:**
   i. In mechanical engineering, components are often rotated to assemble or achieve a desired orientation;
   ii. In robotics, rotations play a vital role in controlling the movement of robotic arms.
2. **Navigation and GPS:** GPS systems use rotations to calculate the position and orientation of vehicles and objects.
3. **Art and Animation:**
   i. Animators use rotations to create lifelike movements in characters and objects;
   ii. Visual artists use rotations to create patterns and symmetrical designs.
4. **Architecture:**
   i. Architects employ rotations to design buildings with interesting and innovative shapes;
   ii. In architectural planning, rotations help visualize building layouts.
5. **Astronomy:**
   i. Astronomers study celestial objects and their rotations to understand the universe better;
   ii. Rotations of planets and moons are critical in celestial mechanics.

Ø **Examples:** Let's explore a few examples to illustrate the concept of rotations:

- **Example 1: Clockwise Rotation:** Consider point A (3, 4) and a rotation of 90° clockwise around the origin. Applying the rotation equations, we find that A' is (–4, 3), indicating a 90° clockwise rotation.
- **Example 2: Counterclockwise Rotation:** Take point B (2, –2) and perform a counterclockwise rotation of 45° around the origin. The new coordinates B' are approximately (0.71, 2.12), signifying a 45° counterclockwise rotation.

- **Example 3: Animation in Games:** In video game development, a character's movement is achieved through rotations. For instance, when a game character jumps, it rotates slightly backward before leaping forward, creating a realistic jumping motion.

Thus, rotations are a fundamental geometric transformation that involves turning a shape around a fixed point by a specified angle. Rotations are crucial for achieving desired orientations and movements in engineering, art, or everyday life. Understanding the mathematical principles behind rotations enables us to model and manipulate objects effectively. Rotations are not just a concept in geometry but a practical tool with wide-ranging applications in the real world, contributing to the functionality and aesthetics of countless objects and systems.

2.5.3. Translations: Sliding Shapes Without Changing Size

Translations are pivotal in various fields, from mathematics and computer graphics to engineering and architecture. In this discussion, we will explore the principles of translations, the mathematical equations that define them, and practical examples showcasing their significance (Anell & Midtkandal, 2017).

2.5.3.1. Principles of Translations

A translation is a transformation that displaces every point of a shape in a consistent direction and distance.

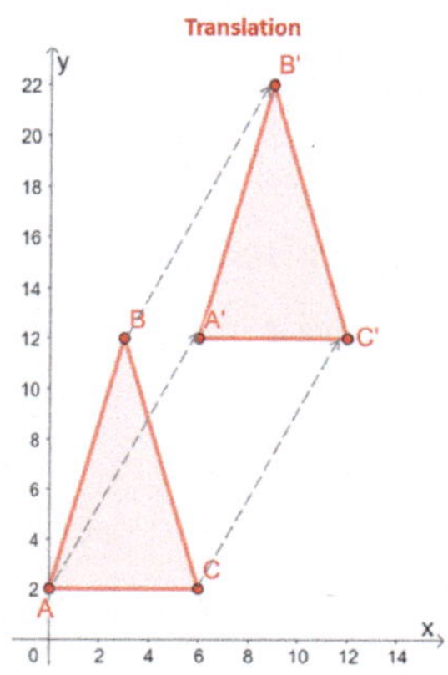

Figure 2.13. Schematic of translation of images.

Source: The Story of Mathematics, Creative Commons License.

Unlike rotations or reflections, translations do not involve turning or flipping the shape; instead, they slide it along a specified path. The entire shape is moved the same distance in the chosen direction, creating a parallel copy of the original (Figure 2.13). In mathematical terms, a translation can be described by specifying the horizontal (x) and vertical (y) distances by which each point of the shape is shifted.

2.5.3.2. Equations for Translations

For a two-dimensional translation, we can express the new coordinates (x,' y') of a point (x, y) as follows:

$$x' = x + a$$

$$y' = y + b$$

Here; (x, y) are the original coordinates; (x,' y') are the translated coordinates; and (a, b) represent the horizontal and vertical distances of the translation, respectively.

For three-dimensional translations, the concept is similar, with the addition of a third coordinate (z):

$$x' = x + a$$

$$y' = y + b$$

$$z' = z + c$$

These equations capture the essence of translation: shifting each point of the shape by the same distances along the x, y, and z axes.

2.5.3.3. Real-World Applications

Translations are widely applied across various domains (Hyde, 2002):

1. **Engineering and Architecture:**
   i. Translations create floor plans in architectural design, moving walls and structures while preserving proportions;
   ii. Engineers use translations to design mechanisms, such as conveyor belts or pistons, which involve precise linear movements.
2. **Computer Graphics and Animation:**
   i. In animation and video games, translations are essential for moving characters and objects smoothly across the screen;
   ii. Graphic design software employs translations to manipulate and arrange images or elements.

3. **Cartography and Geography:**
   i. Cartographers use translations to create maps with different projections and orientations;
   ii. In GPS navigation, translations are applied to track the movement of vehicles or individuals.
4. **Robotics:** Robots rely on translations for precise movement along defined paths, which is crucial in manufacturing and logistics.

Ø **Examples:** Let's explore a few examples to illustrate the concept of translations:

- **Example 1: Sliding a Rectangle:** Consider a rectangle with vertices A (1, 2), B (4, 2), C (4, 5), and D (1, 5). If we apply a translation with distances (2, 3), each point of the rectangle is shifted two units to the right and three units upward. The new coordinates of the vertices become A'(3, 5), B'(6, 5), C'(6, 8), and D'(3, 8).
- **Example 2: Map Projections:** Cartographers use translations to create map projections. For instance, the Mercator projection translates the Earth's curved surface onto a flat map, preserving angles and shapes. This translation is crucial for accurate navigation.
- **Example 3: Robot Arm Movement:** In industrial automation, robotic arms are programmed to perform translations in three-dimensional space. It enables them to pick up objects, move them, and precisely place them in various manufacturing processes.

2.6. SYMMETRY IN SHAPES

Symmetry in shapes refers to a fundamental property where parts of an object or figure exhibit balance or correspondence concerning a specific axis or point. Symmetry plays a vital role in geometry, art, and nature.

It is classified into various types, such as reflection (mirror symmetry), rotational (turning symmetry), and translational symmetry, each with unique characteristics and mathematical principles.

Understanding symmetry enhances our ability to recognize patterns, create aesthetically pleasing designs, and solve complex problems in diverse fields, from mathematics and architecture to biology and crystallography (Ekeh et al., 2021).

2.6.1. Line Symmetry

Line symmetry, also known as reflection symmetry, is a critical concept within the realm of symmetry in shapes. It involves the identification of lines, often referred to as axes of symmetry, along which a shape can be folded or reflected to match its other half.

This concept has significant applications in various fields, from mathematics to art and engineering. In this explanation, we will study the details of line symmetry, exploring how to identify lines of symmetry and examining symmetrical shapes as practical examples (Korchi & Ghanou, 2020).

2.6.1.1. Identifying Lines of Symmetry

To understand line symmetry fully, we first need to grasp how to identify lines of symmetry within different shapes.

A line of symmetry is an imaginary line that divides a shape into two congruent or identical halves. When an object possesses line symmetry, you can fold it along this line, and one half will overlap precisely with the other half. This concept is crucial in recognizing patterns and solving geometric problems (Dogruer & Akyuz, 2020).

Consider the simplest shape: a square. A square has four lines of symmetry, passing through the midpoint of each side. Each of these lines divides the square into two congruent right triangles. In contrast, a rectangle also has lines of symmetry, but they are different from a square. A rectangle has two lines of symmetry, each passing through the midpoint of a longer side and dividing it into two congruent rectangles.

Moving to more complex shapes, cônsider an equilateral triangle. An equilateral triangle has three lines of symmetry, each passing through a vertex and bisecting the opposite side. These lines create three congruent smaller equilateral triangles within the larger one.

2.6.1.2. Symmetrical Shapes

Now, let's explore some common symmetrical shapes that exhibit line symmetry (Buchanan & Collard, 2010):

1. **Circle:** This possesses infinite lines of symmetry. Regardless of where you draw a line passing through the center, it will divide the circle into two congruent halves. It is due to the circle's rotational

symmetry, which means it looks the same when you rotate it by any angle.

2. **Regular Polygons:** Such as equilateral triangles, squares, and hexagons, are highly symmetrical. Equilateral triangles have three lines of symmetry, squares have four, and hexagons have six. These lines pass through specific points in each case, dividing the shapes into congruent parts.
3. **Heart Shape:** It is a classic example of a shape with one line of symmetry. The line passes vertically through the heart's center, creating two mirror-image halves.
4. **Star:** Depending on the number of points on the star, it can have varying degrees of symmetry. For example, a five-pointed star has five lines of symmetry, while an eight-pointed star has eight. These lines intersect at the star's center (Figure 2.14).

In conclusion, line symmetry is a fundamental concept in geometry that helps us understand and identify symmetrical shapes. It involves recognizing lines along which a shape can be folded or reflected to match its other half. By identifying lines of symmetry and studying symmetrical shapes, we gain insights into the inherent order and aesthetics in the geometric world and beyond.

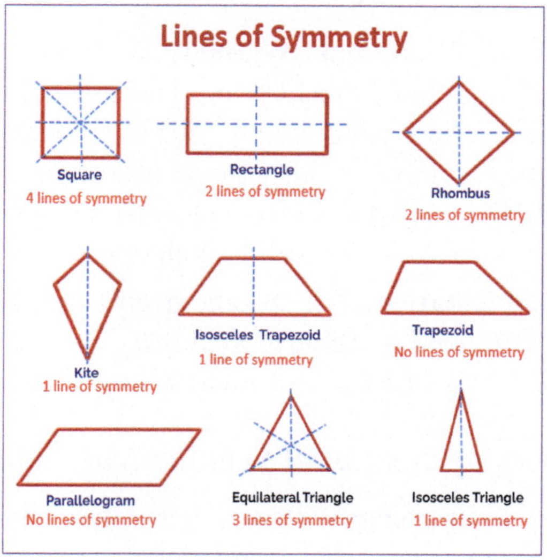

Figure 2.14. Illustration of the lines of symmetry.

Source: Online Math Learning, Creative Commons License.

2.6.2. Rotational Symmetry

Rotational symmetry is another fundamental concept within the study of symmetry in shapes. Unlike line symmetry, which involves folding or reflecting an object, rotational symmetry focuses on the idea that a shape looks the same after a certain degree of rotation around a central point.

In this explanation, we will explore rotational symmetry, including how to recognize it, and provide examples of shapes that exhibit this intriguing property (Borissova et al., 2022).

2.6.2.1. Recognizing Rotational Symmetry

To grasp the concept of rotational symmetry, we need to understand how shapes can be rotated to reveal identical positions. Rotational symmetry occurs when a shape can be rotated by a certain angle (typically less than 360°) around a central point, called the center of rotation, and still appear unchanged or identical in multiple positions.

The minimum angle required for such repetition is the order of rotational symmetry. Let's break It down with some key elements (Boissonnat, 1984):

1. **Center of Rotation:** In any shape exhibiting rotational symmetry, there is a central point around which the rotation occurs. This point serves as the axis or pivot for the rotation.
2. **Order of Rotational Symmetry:** The order of rotational symmetry, often denoted by 'n,' represents the number of times a shape can be rotated within 360° to coincide with its original position. For instance, if a shape has an order of 2, it can be rotated 180° and appear identical, while an order of 3 indicates it can be rotated 120° and still look the same.
3. **Angle of Rotation:** The specific degree measured by which the shape is rotated to achieve symmetry. It is calculated as 360° divided by the order of rotational symmetry ($\theta = 360°/n$).

2.6.2.2. Examples of Rotational Symmetrical Shapes

Now, let's explore some common shapes that exhibit rotational symmetry, providing insight into their order of rotational symmetry and angles of rotation (Pedai et al., 2021):

1. **Circle:** The circle is a prime example of a shape with full rotational symmetry. It has an infinite order of rotational symmetry, as it

looks the same regardless of how much you rotate it around its center. In this case, the rotation angle is $\theta = 360°/\infty = 0°$.

2. **Regular Polygons:** Such as equilateral triangles, squares, pentagons, and hexagons, possess rotational symmetry. The order of rotational symmetry for these shapes corresponds to the number of sides. For instance:
   i. A square has an order of 4, which can be rotated by 90° to coincide with its original position ($\theta = 360°/4 = 90°$);
   ii. An equilateral triangle has an order of 3, requiring a 120° rotation to achieve symmetry ($\theta = 360°/3 = 120°$).
3. **Star:** Depending on the number of points on a star, it can have rotational symmetry. For instance:
   i. A five-pointed star has an order of 5, necessitating a 72° rotation for symmetry ($\theta = 360°/5 = 72°$);
   ii. An eight-pointed star has an order of 8, where each rotation is 45° ($\theta = 360°/8 = 45°$).
4. **Snowflake:** It is a captivating natural example of rotational symmetry. Its intricate, hexagonal structure possesses an order of 6, requiring a 60° rotation for each repetition ($\theta = 360°/6 = 60°$).
5. **Spinning Wheel:** The spokes of a spinning wheel illustrate rotational symmetry. Each spoke can be rotated to align with the original, representing an order of symmetry equal to the number of spokes. For instance, a wheel with 12 spokes has an order of 12, with each spoke requiring a 30° rotation ($\theta = 360°/12 = 30°$).

In conclusion, rotational symmetry is a captivating concept that involves repeating shapes under rotation around a central point. Shapes with rotational symmetry exhibit specific orders of symmetry, each corresponding to a particular rotation angle. Recognizing and understanding rotational symmetry allows us to appreciate the beauty and regularity found in a wide array of geometric and natural forms.

2.7. GEOMETRIC FORMULAS FOR SHAPES

Geometric formulas are essential tools in geometry, providing precise mathematical expressions to calculate various properties of shapes. These formulas enable us to determine key attributes such as area, perimeter, volume, and more. They serve as the mathematical foundation for solving practical problems related to geometric shapes, from finding the area of a room's floor to calculating the volume of complex three-dimensional structures. These formulas simplify complex geometric calculations, making them accessible and applicable in various real-world scenarios, from architecture and engineering to design and construction. Understanding and using these formulas is fundamental for anyone working with geometric shapes, as they empower us to quantify and manipulate the spatial aspects of our physical world (Putra et al., 2021).

2.7.1. Areas Formula

Certainly, understanding how to calculate the area of various two-dimensional shapes is a fundamental aspect of geometry. These formulas allow us to quantify the space enclosed by different geometric figures. This section will explore some common area formulas for 2D shapes and see how they are derived mathematically (Rege et al., 2013).

1. **Area of a Rectangle:** The formula for calculating the area of a rectangle is one of the simplest in geometry:

$$\text{Area} = \text{Length} \times \text{Width}$$

where; the area represents the enclosed space within the rectangle; length is the length of one of the rectangle's sides (usually the longer one); and width is the length of the other side (usually the shorter one).

To visualize It, imagine a rectangular room. You can find its area by measuring the length and width of the floor and multiplying these two values.

2. **Area of a Square:** A square is a special type of rectangle where all sides are of equal length. Therefore, its area formula is:

$$\text{Area} = \text{Side Length} \times \text{Side Length}$$

In this case, both sides have the same length, simplifying the formula.

3. **Area of a Triangle:** The area of a triangle is determined by its base and height. The formula is as follows:

$$\text{Area} = \frac{1}{2} \times \text{Base} \times \text{Height}$$

Here's a brief explanation:

i. The area represents the space enclosed by the triangle;

ii. The base is the length of one of the triangle's sides (usually the bottom);

iii. Height is the length of a perpendicular line dropped from the opposite vertex to the base.

This formula is particularly useful for finding the area of various triangular objects, such as rooftops or land plots (Shin et al., 2015).

4. **Area of a Circle:** The area of a circle is calculated using the mathematical constant π (pi). The formula is:

$$\text{Area} = \pi \times \text{Radius}^2$$

Here's what each element represents:

i. The area is the space enclosed by the circle;

ii. π (pi) is approximately equal to 3.14159 and is a fundamental constant in geometry;

iii. Radius is the distance from the circle's center to any point on its edge.

This formula is essential for determining the area of circular objects like wheels, plates, or circular fields.

5. **Area of a Parallelogram:** A parallelogram's area is determined by its base and height, similar to a triangle:

$$\text{Area} = \text{Base} \times \text{Height}$$

The key difference is that for a parallelogram, the height is the perpendicular distance between the two parallel sides.

These are some of the fundamental area formulas for common two-dimensional shapes. Understanding and applying these formulas is crucial for solving real-world area calculation problems. Whether designing a garden, building a room, or working in various fields like architecture, engineering, or art, these formulas provide the mathematical tools to quantify space accurately (Tangelder & Veltkamp, 2008).

2.7.2. Perimeter Formulas

Here, we will explore the common perimeter formulas for various 2D shapes and how they are derived mathematically (Yan et al., 2016).

1. **Perimeter of a Rectangle:** The formula for finding the perimeter of a rectangle is straightforward:

$$\text{Perimeter} = 2 \times (\text{Length} + \text{Width})$$

where; perimeter represents the total length around the rectangle; length is one of the longer sides; and width is one of the shorter sides.

To visualize it, consider a rectangular fence. To calculate the fencing required, you'd use the above formula by summing up the lengths of all four sides.

2. **The Perimeter of a Square:** For a square where all sides are equal in length, the perimeter formula is:

   $$\text{Perimeter} = 4 \times \text{Side Length}$$

   In this case, you multiply the length of one side by 4, as all sides are identical.

3. **The Perimeter of a Triangle:** The perimeter of a triangle is found by adding the lengths of its three sides:

   $$\text{Perimeter} = \text{Side}_1 + \text{Side}_2 + \text{Side}_3$$

   The above formula applies to any triangle, whether it's equilateral (all sides equal), isosceles (two sides equal), or scalene (no sides equal) (Zerroug & Nevatia, 1993).

4. **Perimeter of a Circle:** To calculate the perimeter (more commonly known as the circumference) of a circle, we use the formula:

   $$\text{Perimeter (Circumference)} = 2 \times \pi \times \text{Radius}$$

Here, π (pi) is approximately equal to 3.14159, and the radius is the distance from the circle's center to its edge. This formula is essential for determining the length of circular objects such as wheels or the borders of circular fields.

5. **The Perimeter of a Parallelogram:** For a parallelogram, the perimeter is found by adding the lengths of all four sides:

   $$\text{Perimeter} = 2 \times \left(\text{Side}_1 + \text{Side}_2\right)$$

In this formula, we take advantage of the opposite sides of a parallelogram having equal lengths.

These are some fundamental perimeter formulas for common two-dimensional shapes.

Understanding and applying these formulas is vital for real-world applications such as construction, fencing, art, and navigation. Whether building a room, planning a garden, or designing a piece of artwork, these formulas provide the mathematical tools to accurately calculate the amount of material or boundary needed (Zhong et al., 2019).

2.7.3. Volume Formulas

Certainly, calculating the volume of three-dimensional shapes is crucial in geometry, as it helps us understand the amount of space enclosed by these shapes. Here, we'll explore the volume formulas for common 3D shapes and how they are mathematically derived.

1. **Volume of a Cube:** A cube is a three-dimensional shape with all sides of equal length. To calculate its volume, you can use the formula:

$$\text{Volume} = \text{Side Length}^3$$

Here, the side length represents the length of one cube's edge. The formula derives from the fact that a cube has six identical square faces, and each face has an area of side length squared (Side Length × Side Length). Therefore, to find the volume, you multiply this area by the side length.

2. **Volume of a Rectangular Prism:** For a rectangular prism, also known as a rectangular box, the formula for volume is:

$$\text{Volume} = \text{Length} \times \text{Width} \times \text{Height}$$

In this formula, the length, width, and height represent the three perpendicular dimensions of the prism. To visualize It, think of a shoebox. To calculate its volume, multiply its length, width, and height.

3. **Volume of a Cylinder:** A cylinder consists of two circular bases and a curved surface. To calculate its volume, use the formula:

$$\text{Volume} = \text{Length} \times \text{Width} \times \text{Height}$$

In this formula, π (pi) is approximately equal to 3.14159, the radius is the distance from the center of one base to the edge, and the height is the distance between the two bases. This formula is essential for determining the volume of objects like soda cans or storage tanks (Yan et al., 2016).

4. **Volume of a Sphere:** A sphere is a perfectly round 3D shape. To find its volume, you can use the formula:

$$\text{Volume} = \frac{4}{3}\pi \times \text{Radius}^3$$

The above formula derives from calculus and integration but is widely used for simplicity. It relates the sphere's radius to its volume, emphasizing the cube of the radius in the calculation.

5. **Volume of a Cone:** A cone has a circular base and a single curved surface that meets at the apex. To calculate its volume, you can use the formula:

$$\text{Volume} = \frac{1}{3}\pi \times \text{Radius}^2 \times \text{Height}$$

Here, the radius is the radius of the circular base, and the height is the distance from the base to the apex. This formula is commonly used in geometry and engineering, particularly for objects like traffic cones and ice cream cones. These volume formulas are fundamental in various fields, including engineering, architecture, physics, and more. They enable us to quantify the amount of space occupied by objects in the real world, aiding in design, construction, and analysis.

Understanding these formulas is crucial for solving practical problems related to volume, whether designing a container, calculating the capacity of a tank, or determining the volume of a room for construction purposes.

2.7.4. Surface Area Formulas

Certainly, understanding surface area formulas for three-dimensional shapes is crucial in geometry as it allows us to quantify the amount of material needed to cover or paint an object, which is essential in various real-world applications. Let's delve into the surface area formulas for common 3D shapes and how they are calculated mathematically (Shin et al., 2015).

1. **Surface Area of a Cube:** A cube has six identical square faces. To calculate its total surface area, you can use the formula:

   $$\text{Surface Area} = 6 \times (\text{Side Length})^2$$

Here, the side length represents the length of one cube's edge. The formula comes from the fact that each face has an area of side length squared, with six identical faces.

2. **Surface Area of a Rectangular Prism:** For a rectangular prism (a rectangular box), the surface area formula is:

   $$\text{Surface Area} = 2(\text{Length} \times \text{Width} + \text{Width} \times \text{Height} + \text{Height} \times \text{Length})$$

This formula accounts for each of the six faces of the rectangular prism. It sums the areas of the top and bottom faces (Length × Width twice), the front and back faces (Width × Height twice), and the two side faces (Height × Length twice).

3. **Surface Area of a Cylinder:** A cylinder consists of two circular bases and a curved surface. To calculate its surface area, use the formula:

$$\text{Surface Area} = 2\pi\text{Radius}^2 + 2\pi\text{Radius} \times \text{Height}$$

In this formula, π (pi) is approximately equal to 3.14159, the radius is the distance from the center of one base to the edge, and the height is the distance between the two bases. The formula sums the areas of both circular bases and the lateral surface area, which is a rectangle rolled into a cylinder shape.

4. **Surface Area of a Sphere:** A perfectly round 3D shape. To find its surface area, you can use the formula:

$$\text{Surface Area} = 4\pi\text{Radius}^2$$

The above formula accounts for the entire surface area of the sphere. It emphasizes the role of the radius, as it is squared in the calculation (Putra et al., 2021).

5. **Surface Area of a Cone:** A cone has a circular base and a single curved surface that meets at a point called the apex. To calculate its surface area, you can use the formula:

$$\text{Surface Area} = \pi\text{Radius} \times (\text{Radius} + \text{Slant Height})$$

Here, the radius is the radius of the circular base, and the slant height is the distance from the apex to any point on the circumference of the base. The formula combines the areas of the circular base and the lateral surface area, a sector of a circle unrolled into a flat shape. These surface area formulas are fundamental for various applications, including architecture, packaging, and manufacturing. They help us determine the amount of material needed to construct objects or cover their surfaces, making them valuable tools for design and construction. Whether planning to paint a cylindrical tank, wrap a gift box, or design a decorative ornament, understanding these surface area formulas is essential for practical problem-solving (Pizlo et al., 2010).

Activity 2.1

Imagine you are an interior designer tasked with arranging furniture in a rectangular room. How would your understanding of geometric shapes, especially rectangles and their properties, help you optimize the furniture layout for both aesthetics and functionality?

2.8. SUMMARY

This chapter introduces the basics of shapes. It explains the importance of understanding shapes and classifies them into two-dimensional, three-dimensional, and irregular shapes. The chapter covers properties of geometric shapes, such as dimensions and properties of polygons. It also delves into congruence and similarity among shapes, geometric transformations like reflections and rotations, symmetry types, and various geometric formulas for areas, perimeters, volumes, and surface areas. This chapter provides foundational knowledge about shapes and their characteristics, setting the stage for more advanced geometric concepts.

REFERENCES

1. Alexiadis, A., (2006). 2D radiation field in photocatalytic channels of square, rectangular, equilateral triangular, and isosceles triangular sections. *Chemical Engineering Science*, *61*(2), 516–525.
2. Anell, I., & Midtkandal, I., (2017). The quantifiable clinothem – types, shapes, and geometric relationships in the Plio-Pleistocene giant forests formation, Taranaki basin, New Zealand. *Basin Research*, *29*, 277–297.
3. Boissonnat, J. D., (1984). Geometric structures for three-dimensional shape representation. *ACM Transactions on Graphics (TOG)*, *3*(4), 266–286.
4. Borissova, D., Ivanova, T., Buhtiarov, N., Naidenov, N., Rasheva-Yordanova, K., Yoshinov, R., & Garvanov, I., (2022). Application of information technology in the teaching of mathematics when study of 2D geometric shapes. In: *2022 45th Jubilee International Convention on Information, Communication, and Electronic Technology (MIPRO)* (Vol. 2, pp. 638–643). IEEE.
5. Brechbühler, C., Gerig, G., & Kübler, O., (1995). Parametrization of closed surfaces for 3D shape description. *Computer Vision and Image Understanding*, *61*(2), 154–170.
6. Buchanan, B., & Collard, M., (2010). A geometric morphometrics-based assessment of blade shape differences among Paleoindian projectile point types from western North America. *Journal of Archaeological Science*, *37*(2), 350–359.
7. Dogruer, S. S., & Akyuz, D., (2020). Mathematical practices of eighth graders about 3D shapes in an argumentation, technology, and design-based classroom environment. *International Journal of Science and Mathematics Education*, *18*, 1485–1505.
8. Drake, A. G., (2011). Dispelling dog dogma: An investigation of heterochrony in dogs using 3D geometric morphometric analysis of skull shape. *Evolution and Development*, *13*(2), 204–213.
9. Ekeh, M. C., Venketsamy, R., Unamba, E. C., & Ugochukwu, N. J., (2021). Using blended-learning strategy in mathematics to strengthen the teaching of geometric 2D shapes and their properties to primary 6 learners. *The Independent Journal of Teaching and Learning*, *16*(2), 85–100.

10. El Korchi, A., & Ghanou, Y., (2020). 2D geometric shapes dataset – for machine learning and pattern recognition. *Data in Brief*, *32*, 106090.
11. Elia, I., & Gagatsis, A., (2003). Young children's understanding of geometric shapes: The role of geometric models. *European Early Childhood Education Research Journal*, *11*(2), 43–61.
12. Huang, W. M., Zhao, Y., Wang, C. C., Ding, Z., Purnawali, H., Tang, C., & Zhang, J. L., (2012). Thermo/chemo-responsive shape memory effect in polymers: A sketch of working mechanisms, fundamentals, and optimization. *Journal of Polymer Research*, *19*, 1–34.
13. Hyde, C., (2002). The electromagnetic response of a sphere in the moment domain: A building block for the construction of arbitrary 3D shapes. In: *SEG Technical Program Expanded Abstracts 2002* (Vol. 2, pp. 645–648). Society of Exploration Geophysicists.
14. İbili, E., Çat, M., Resnyansky, D., Şahin, S., & Billinghurst, M., (2020). An assessment of geometry teaching supported with augmented reality teaching materials to enhance students' 3D geometry thinking skills. *International Journal of Mathematical Education in Science and Technology*, *51*(2), 224–246.
15. Iwamoto, M., & Ou-Yang, Z. C., (2004). Shape deformation and circle instability in two-dimensional lipid domains by dipolar force: A shape-and size-dependent line tension model. *Physical Review Letters*, *93*(20), 206101.
16. Kimia, B. B., Tannenbaum, A. R., & Zucker, S. W., (1995). Shapes, shocks, and deformations I: The components of two-dimensional shape and the reaction-diffusion space. *International Journal of Computer Vision*, *15*, 189–224.
17. Kirner, T. G., Reis, F. M. V., & Kirner, C., (2012). Development of an interactive book with augmented reality for teaching and learning geometric shapes. In: *7th Iberian Conference on Information Systems and Technologies (CISTI 2012)* (Vol. 2, pp. 1–6). IEEE.
18. Kuang, W., Hou, Z., & Liu, Y., (2004). The effects of shapes and symmetries of scatterers on the phononic band gap in 2D phononic crystals. *Physics Letters A*, *332*(5–6), 481–490.
19. Low, T. B., & List, R., (1982). Collision, coalescence, and breakup of raindrops. Part I: Experimentally established coalescence efficiencies and fragment size distributions in breakup. *Journal of Atmospheric Sciences*, *39*(7), 1591–1606.

20. Markov, A., Trofimov, A., & Sevostianov, I., (2020). A unified methodology for calculation of compliance and stiffness contribution tensors of inhomogeneities of arbitrary 2D and 3D shapes embedded in isotropic matrix – open access software. *International Journal of Engineering Science*, *157*, 103390.
21. Ngom, N. F., Monga, O., Mohamed, M. M. O., & Garnier, P., (2012). 3D shape extraction segmentation and representation of soil microstructures using generalized cylinders. *Computers & Geosciences*, *39*, 50–63.
22. Pedai, S. S., Sulistiawati, S., & Arifin, S., (2021). The identification of students' mistakes on mathematical communication ability in three-dimensional shapes of geometry: Cube and cuboid. In: *AIP Conference Proceedings*. AIP Publishing. (Vol. 2331, pp. 231–233).
23. Pizlo, Z., Sawada, T., Li, Y., Kropatsch, W. G., & Steinman, R. M., (2010). New approach to the perception of 3D shape based on veridicality, complexity, symmetry, and volume. *Vision Research*, *50*(1), 1–11.
24. Putra, Z. H., Panjaitan, I. O. D., Putri, N. A., Wulandari, T. R., & Hermita, N., (2021). Design and implementation of GeoGebra learning activities of area and perimeter of rectangles for primary school students. In: *Journal of Physics: Conference Series* (Vol. 2049, pp. 12032–12035). IOP Publishing.
25. Rege, S., Memane, R., Phatak, M., & Agarwal, P., (2013). 2D geometric shape and color recognition using digital image processing. *International Journal of Advanced Research in Electrical, Electronics, and Instrumentation Engineering*, *2*(6), 2479–2487.
26. Shin, D. H., Yun, K. N., Jeon, S. G., Kim, J. I., Saito, Y., Milne, W. I., & Lee, C. J., (2015). High performance field emission of carbon nanotube film emitters with a triangular shape. *Carbon*, *89*, 404–410.
27. Tangelder, J. W., & Veltkamp, R. C., (2008). A survey of content-based 3D shape retrieval methods. *Multimedia Tools and Applications*, *39*, 441–471.
28. Wu, Y., Ojha, A., Patriarca, L., & Sehitoglu, H., (2015). Fatigue crack growth fundamentals in shape memory alloys. *Shape Memory and Superelasticity*, *1*, 18–40.
29. Yan, X., Yang, J., Yumer, E., Guo, Y., & Lee, H., (2016). Perspective transformer nets: Learning single-view 3D object reconstruction without 3D supervision. *Advances in Neural Information Processing Systems*, *29*, 2524–2526.

30. Zerroug, M., & Nevatia, R., (1993). Quasi-invariant properties and 3-D shape recovery of non-straight, non-constant generalized cylinders. In: *Proceedings of IEEE Conference on Computer Vision and Pattern Recognition* (Vol. 2, pp. 96–103). IEEE.
31. Zhong, X., Chen, F., Nie, H., Wang, R., & Xu, Z., (2019). Reducing the energy consumption and cell sludge of the zinc electrowinning process by using a pyramid-shaped 3D-Pb anode. *Hydrometallurgy, 190*, 105188.

CHAPTER 3

Spatial Relationships and Transformations

Contents

3.1. INTRODUCTION

The study of spatial relationships holds a central position in geometry. Spatial relationships concern the interactions and arrangements of objects and points within three-dimensional space. In this chapter, we will study the fundamental concepts of spatial relationships, providing a clear understanding of their definition, significance, and practical applications (Kučerová et al., 2015). Spatial relationships encompass how objects and points can be positioned relative to each other in three-dimensional space. It involves examining the distances, angles, and orientations between objects and understanding how these factors influence their interactions. These relationships are the backbone of geometry in three dimensions, providing the foundation for geometric analysis and problem-solving in theoretical and real-world contexts (Harris et al., 2013).

The study of spatial relationships is of paramount importance for several reasons:

1. **Real-World Scenarios:** Spatial relationships are abundant in our daily lives. Whether it is the arrangement of buildings in a city, the design of intricate machinery, or the layout of molecules in chemistry, understanding spatial relationships is essential for comprehending and optimizing these real-world scenarios.
2. **Engineering and Design:** Engineers and designers rely heavily on spatial relationships to create efficient and functional structures and systems. From architects planning the layout of a building to aerospace engineers designing the shape of an aircraft, spatial relationships play a critical role in achieving desired outcomes.
3. **Problem Solving:** Spatial geometry provides a powerful framework for solving complex problems. It aids in analyzing spatial data, making predictions, and optimizing solutions. For instance, understanding spatial relationships in logistics can help minimize travel distances and reduce costs.
4. **Science and Research:** In fields such as physics, chemistry, and biology, spatial relationships are central to understanding the behavior and interactions of physical entities. Researchers use spatial models to predict how particles interact, how molecules fold, or how celestial bodies move in space (Cantù, 2010).
5. **Visualization Skills:** Developing spatial visualization skills, honed through studying spatial relationships, is crucial in

various professions. Surgeons, for example, rely on their ability to manipulate three-dimensional structures when performing delicate procedures mentally.

6. **Innovation and Technology:** Many technological advancements, from virtual reality to 3D printing, leverage our understanding of spatial relationships. Innovations in these fields are often grounded in the principles of spatial geometry.
7. **Educational Foundations:** Spatial relationships are introduced in early education to foster essential cognitive skills. This early exposure lays the groundwork for more advanced mathematical concepts in later education.

In conclusion, spatial relationships are the basis of geometry in three dimensions. They find applications in countless fields, from engineering and design to problem-solving and scientific research. This chapter will equip you with the necessary tools to grasp and apply these fundamental concepts effectively, allowing you to explore the fascinating world of spatial geometry (Marino et al., 2008).

3.2. DISTANCE AND LENGTH

Euclidean distance is a fundamental concept in geometry, especially when dealing with spatial relationships. It provides a straightforward and intuitive way to measure the length or distance between two points in two-dimensional (2D) and three-dimensional (3D) space. Named after the ancient Greek mathematician Euclid, this distance metric is based on the Pythagorean theorem and forms the basis of Euclidean geometry. This section will study Euclidean distance in 2D and 3D, examining its mathematical representation, properties, and practical applications.

3.2.1. Euclidean Distance in Two Dimensions (2D)

Let's start by considering the Euclidean distance between A and B points in a 2D plane. These points can be represented as coordinates: A (x_1, y_1) and B (x_2, y_2). The Euclidean distance, often denoted as 'd,' between A and B is given by the Pythagorean theorem:

$$d = \sqrt{(x_2 - x_1)^2 + (y_2 - y_1)^2}$$

Here's a breakdown of the components; x_1 and x_2 represent the x-coordinates of points A and B, respectively; y_1 and y_2 represent the y-coordinates of points A and B, respectively; the expression $(x_2 - x_1)^2$ calculates the squared horizontal distance between the two points; and similarly, $(y_2 - y_1)^2$ calculates the squared vertical distance between the two points.

Taking the square root of the sum of these squared distances gives the Euclidean distance.

Let's illustrate this with an example:

Ø **Example:** Find the Euclidean distance between points A (2, 3) and B (6, 8) in a 2D plane.

Using the formula:

$$d = \sqrt{(6-2)^2 + (8-3)^2} = \sqrt{4^2 + 5^2} = \sqrt{41}$$

So, the Euclidean distance between points A and B is $(\sqrt{41})$ units.

Following are the properties of the Euclidean Distance in 2D:

1. **Non-Negativity:** The Euclidean distance is always non-negative. It is zero if and only if the two points are coincident.
2. **Symmetry:** The distance between points A and B is the same as between points A and A.
3. **Triangle Inequality:** For any three points A, B, and C, the sum of the distances from A to B and B to C is always greater than or equal to the distance from A to C (Trafton et al., 2006).

3.2.2. Euclidean Distance in Three Dimensions (3D)

Extending the concept to three dimensions involves adding the z-coordinate to our coordinate representation. Consider two points A (x_1, y_1, z_1) and B (x_2, y_2, z_2) in 3D space. The Euclidean distance 'd' between these points can be calculated using the 3D Pythagorean theorem:

$$d = \sqrt{(x_2 - x_1)^2 + (y_2 - y_1)^2 + (z_2 - z_1)^2}$$

In his 3D formula, all components are analogous to the 2D case, with the addition of z-coordinates to account for the third dimension. Let's see this in action:

Ø **Example:** Find the Euclidean distance between points A (1, 2, 3) and B (4, 6, 8) in 3D space.

Using the 3D formula:

$$d = \sqrt{(4-1)^2 + (6-2)^2 + (8-3)^2} = \sqrt{3^2 + 4^2 + 5^2} = \sqrt{50}$$

So, the Euclidean distance between points A and B in 3D is $\left(\sqrt{50}\right)$ units.

The properties of Euclidean distance in 3D are consistent with those in 2D, including non-negativity, symmetry, and triangle inequality.

3.2.3. Applications in Real-Life Scenarios

The distance formula, a fundamental tool in geometry, is not just a mathematical concept confined to textbooks and classrooms. Its practical applications extend into various real-life scenarios, where precise measurement of distances is critical. This section explores how the distance formula is used in the real world (Paul et al., 2012).

3.2.3.1. Navigation and GPS Technology

The widespread use of GPS (Global Positioning System) technology in our daily lives relies heavily on the distance formula. GPS devices determine our location and provide directions by calculating the distances between our current coordinates and our desired destination. GPS receivers receive signals from multiple satellites orbiting the Earth. By measuring the time, it takes for signals to travel from the satellites to the receiver, the GPS unit can determine the distance to each satellite. The device then uses trilateration, a technique based on the distance formula, to pinpoint the receiver's location. This information is invaluable for applications such as driving directions, emergency services, and location-based smartphone services (Shelton & Hedley, 2004).

Ø **Example: Haversine Formula for Distance Calculation**

- **Scenario:** You want to calculate the distance (in kilometers) between two GPS coordinates (φ1, λ1) and (φ2, λ2) on the Earth's surface.
- **Mathematical Equation (Haversine Formula):**

$$a = \sin^2(\Delta\varphi/2) + \cos(\varphi 1) * \cos(\varphi 2) * \sin^2(\Delta\lambda/2)$$

$$c = 2 * atan2\left(\sqrt{a}, \sqrt{(1-a)}\right)$$

$$d = R * c$$

where; Δφ is the difference in latitude between the two points in radians; Δλ is the difference in longitude between the two points in radians; φ1 and φ2 are the latitudes of the two points in radians; R is the Earth's radius (mean radius = 6,371 kilometers); and d is the distance between the two points in kilometers.

3.2.3.2. Architectural Design and Construction

In architecture and construction, precise measurements of distances are essential to ensure that buildings, bridges, and structures are designed and constructed accurately. Architects and engineers use the distance formula to calculate dimensions, establish proportions, and determine the placement of structural elements (Zhu et al., 2022). For example, when designing a building, architects use the distance formula to calculate the length of beams, the height of walls, and the spacing between columns. Similarly, civil engineers use it to lay out roadways, bridges, and tunnels.

Ø **Example: Calculating the Length of a Diagonal in a Rectangular Room:**

- **Problem:** You have a rectangular room where one side (a) measures 5 meters, and the other (b) measures 8 meters. Find the length of the diagonal (D) of the room.
- **Solution:**

1. Identify the known values:

i. Length of one side (a) = 5 meters;

ii. Length of the other side (b) = 8 meters.

2. Use the Pythagorean Theorem to calculate the length of the diagonal (D):

$$D = \sqrt{\left(a^2 + b^2\right)}$$

3. Substitute the known values into the equation:

$$D = \sqrt{(5^2 + 8^2)}$$

4. Calculate the squares of the side lengths:

$$D = \sqrt{(25 + 64)}$$

5. Add the squares together:

$$D = \sqrt{(89)}$$

6. Calculate the square root of 89:

D ≈ 9.43 meters (rounded to two decimal places)

The length of the diagonal (D) of the rectangular room is approximately 9.43 meters (Qiao et al., 2018).

3.2.3.3. Surveying and Land Measurement

Land surveyors use the distance formula extensively to measure distances and map out plots of land. Surveyors use accurate distance calculations to determine property boundaries, conduct topographic surveys, or create land-use plans.

Surveying instruments, such as total stations and theodolites, calculate distances between survey points using the distance formula. This information is crucial for land development, property disputes, and urban planning (Bryant, 1998).

- Ø **Example: Calculating Land Area with a Theodolite and Chain Survey:**
- **Scenario:** You need to measure the area of an irregularly shaped piece of land using a theodolite and chain survey method.
- **Solution:**

1. **Linear Measurements (Chain Survey):** Measure the lengths of sides of the land parcel using a chain or tape measure and record these lengths.
2. **Angle Measurements (Theodolite):** Measure the internal angles at the corners of the land parcel using theodolite. Record these angles.
3. **Area Calculation (Trigonometry):** Calculate the area using the following formula:

$$Area(A) = 0.5 \times \Sigma\left(Length_i \times Length_{(i+1)} \times sin(Angle_i)\right)$$

where; $Length\_i$ and $Length_{(i+1)}$ are consecutive side lengths; $Angle_i$ is the included angle between the side $Length_i$ and $Length_{(i+1)}$; and Σ represents the summation over all sides of the land parcel. Suppose you have measured the following data for an irregular land parcel:

- **Side Lengths:**

  $Length_1 = 50\,meters$

  $Length_2 = 75\,meters$

  $Length_3 = 60\,meters$

  $Length_4 = 45\,meters$

- Internal Angles (Measured Counterclockwise):

  $Angle_1 = 100\,degrees$

  $Angle_2 = 75\,degrees$

  $Angle_3 = 110\,degrees$

  $Angle_4 = 85\,degrees$

You can calculate the area of the land parcel using the provided data:

$$Area(A) = 0.5 \times \left[(50 \times 75 \times sin(100)) + (75 \times 60 \times sin(75)) + (60 \times 45 \times sin(110)) + (45 \times 50 \times sin(85))\right]$$

Perform the calculations for each term and sum them up:

Area (A) $\approx$ 3,750 square meters

The area of the irregular land parcel is approximately 3,750 square meters (Sorrows & Hirtle, 1999).

3.2.3.4. Astronomy and Celestial Navigation

In astronomy and celestial navigation, distances between celestial objects and Earth are calculated using variations of the distance formula. Astronomers use it to determine the distances to stars, galaxies, and other celestial bodies (Koya, 2017).

Ø **Example:** The parallax method relies on the Earth's orbit around the Sun to measure distance to nearby stars. By observing a star's apparent shift in position against the background of more distant stars at different times of the year, astronomers can calculate its distance using trigonometry, which ultimately involves the distance formula (Figure 3.1).

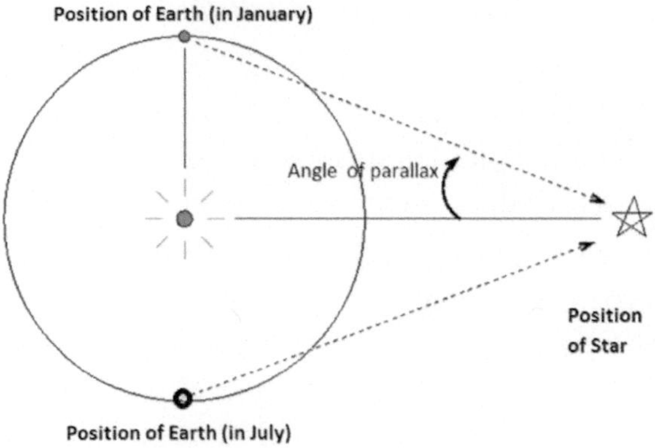

Figure 3.1. Illustration of parallax method for measuring the distance of a near star.

Source: Vedantu, Creative Commons License.

3.2.3.5. Sports and Athletic Performance Analysis

The distance formula has applications in sports and athletic performance analysis. It is often used in tracking and measuring the distances athletes cover in various sports, such as running, cycling, and swimming (Schulz et al., 2002).

Sports scientists and coaches use GPS devices and accelerometers to collect data on an athlete's movement. By analyzing this data, they can calculate distances traveled, speeds achieved, and energy expenditure. This information is crucial for optimizing training, injury prevention, and enhancing performance.

3.2.3.6. Environmental Monitoring and Ecology

Ecologists and environmental scientists use the distance formula to study and monitor ecosystems and wildlife. For instance, when tracking the movement patterns of animals, researchers attach GPS collars or tags to animals. These devices record location data and then calculate distances between the animal's positions at different times (Solari et al., 2007).

Understanding the distances animals travel and the areas they inhabit is essential for conservation efforts, habitat preservation, and wildlife management.

3.2.3.7. Robotics and Autonomous Vehicles

Robotics and autonomous vehicles rely on the distance formula for navigation and obstacle avoidance. Robots with sensors, such as LiDAR (Light Detection and Ranging), ultrasonic sensors, and cameras, use distance calculations to perceive their surroundings (Nabil et al., 1995).

Autonomous cars, drones, and robots utilize the distance formula to determine the proximity of objects and obstacles. This information is crucial for making real-time decisions on path planning and collision avoidance (Figure 3.2).

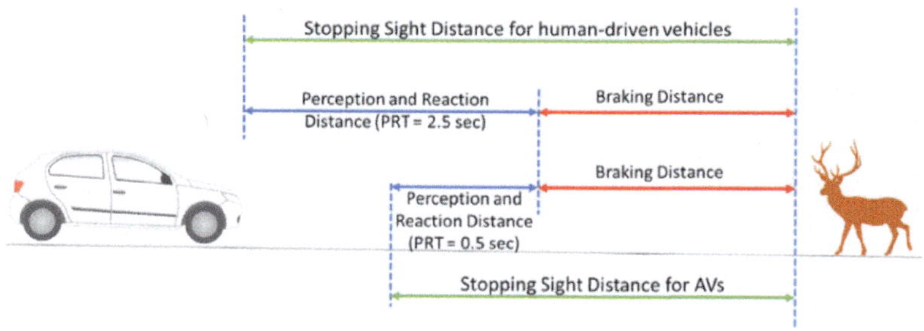

Figure 3.2. Schematic of autonomous car that uses distance formula.

Source: Kareem Othman, Creative Commons License.

Hence, the distance formula is a mathematical foundation for accurately measuring distances and spatial relationships in these real-life scenarios. Its practical applications span diverse fields, from technology and engineering to astronomy and ecology, making it an indispensable tool in our modern world. Understanding and applying the distance formula enhances our ability to navigate, design, measure, and analyze our environment's physical and spatial aspects.

3.3. MIDPOINT AND BISECTORS

Midpoints and bisectors play a crucial role in geometry. The midpoint of a line segment is the point exactly halfway between its two endpoints. It is calculated as the average of the coordinates of the endpoints, making it a fundamental concept in coordinate geometry.

Conversely, bisectors are lines, rays, or line segments that divide an angle or a line segment into two equal parts. Angle bisectors split an angle into two congruent angles, while segment bisectors divide a line segment into two equal parts. These concepts are essential for understanding symmetry, congruence, and various geometric constructions, making them fundamental in studying geometry.

3.3.1. Midpoint Formula

The midpoint allows us to find the exact center of a line segment, dividing it into two equal parts. This concept is essential for understanding geometry and finds applications in various real-world scenarios, from navigation to computer graphics (Liao, 2017).

Tip

When you need to find the midpoint between two points in a coordinate system, you can use the midpoint formula, which averages the x and y coordinates separately. It is especially handy for tasks like splitting a line segment in half.

Mathematically, if we have two endpoints, A (x_1, y_1) and B (x_2, y_2), the coordinates of the midpoint M can be found using the midpoint formula:

$$\mathrm{M} = \left(\frac{x_1 + x_2}{2}, \frac{y_1 + y_2}{2} \right)$$

In this formula, (x_1, y_1) and (x_2, y_2) Represent the coordinates of the two endpoints of the line segment AB and (x, y) represents the coordinates of the midpoint M.

Ø **Example: Finding the Midpoint**

We have two endpoints, A (3, 4) and B (7, 10). To find the midpoint M, we use the midpoint formula:

$$M = \left(\frac{3+7}{2}, \frac{4+10}{2} \right)$$

Simplifying this expression, we get:

$$M = \left(\frac{10}{2}, \frac{14}{2} \right)$$

$$M = (5, 7)$$

So, the midpoint of line segment AB is M (5, 7).

3.3.1.1. Properties of the Midpoint

Understanding the properties of the midpoint formula is essential in geometry. Here are some key properties:

1. **Symmetry:** The midpoint divides the line segment into two equal parts, making it symmetric concerning the midpoint. This property is crucial for understanding congruence and symmetry in geometric shapes (David, 2012).
2. **Distance:** The distance between each endpoint and the midpoint is equal. This property is significant in understanding geometric proofs and theorems.
3. **Coordinates:** The midpoint coordinates are the average of the coordinates of the endpoints. This property is essential in coordinate geometry, where the midpoint formula is frequently used.

3.3.1.2. Visualizing the Midpoint

Ø **Example:** Consider line segment AB, where A (3, 4) and B (7, 10). Imagine drawing this line segment on a coordinate plane to visualize the midpoint.

1. Locate point A (3, 4) on the coordinate plane.
2. Locate point B (7, 10) on the coordinate plane.
3. Draw the line segment AB connecting points A and B.
4. Use the midpoint formula to calculate the midpoint M (5, 7).
5. Locate point M (5, 7) on the coordinate plane.

You will notice that point M is exactly in the middle of line segment AB, equidistant from both endpoints. It demonstrates the practical application of the midpoint formula in geometry (Demetriou, 2006).

3.3.2. Angle Bisectors and Perpendicular Bisectors

Angle and perpendicular bisectors are essential tools in geometry, playing a crucial role in dividing and connecting geometric elements. These concepts provide a deeper understanding of angles, segments, and shapes, and their properties are central to various geometric proofs and constructions (Shu et al., 2023).

3.3.2.1. Angle Bisectors

An angle bisector is a line or ray that divides an angle into two equal parts, each with the same measure. In other words, it cuts the angle into two congruent angles. Angle bisectors are often denoted by a straight line with two arrows pointing away from the angle's vertex (∠ABC) (Figure 3.3).

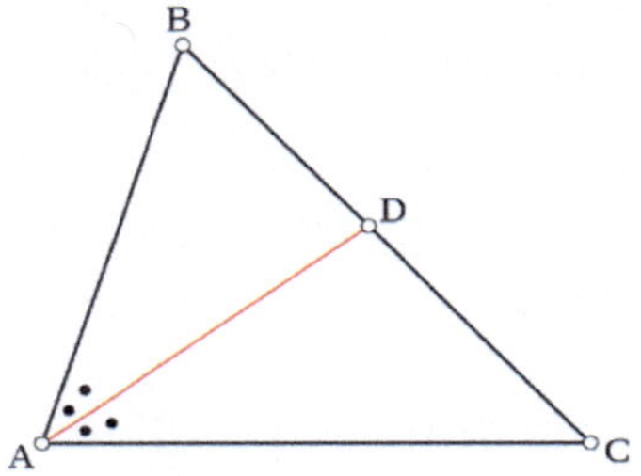

Figure 3.3. Schematic of the angle bisector.

Source: Limaner, Creative Commons License.

The following are the properties of the angle bisector:

1. **Equal Angles:** An angle bisector divides an angle into two angles with equal measures. If ∠ABC is bisected by line AD, then ∠BAD = ∠DAC.
2. **Vertex on Bisector:** The vertex of the angle being bisected lies on the angle bisector. In the example above, vertex A lies on line AD (Heuel & Forstner, 2001).

3.3.2.2. Perpendicular Bisectors

A perpendicular bisector is a line, ray, or segment that divides a line segment into two equal parts and is perpendicular to the line segment at its midpoint. It is often denoted by a straight line with a small box symbol (||) (Figure 3.4).

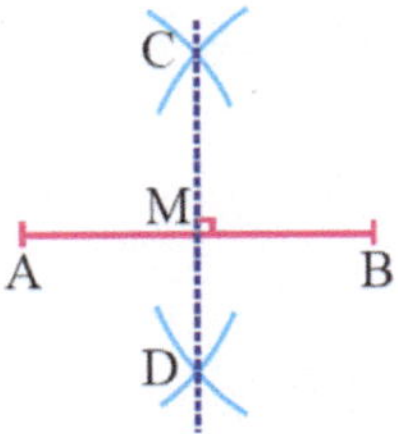

Figure 3.4. Schematic of the perpendicular bisector.

Source: Online Math 4 all, Creative Commons License.

The following are the properties of the perpendicular bisector:

1. **Equal Segments:** A perpendicular bisector divides a line segment into two equal parts. If AB is bisected by line CD at point M, then AM = MB.
2. **Perpendicular Lines:** The perpendicular bisector is perpendicular to the line segment it bisects. In the example above, line CD is perpendicular to segment AB (Gleiss et al., 2011).
3. **Midpoint:** The perpendicular bisector passes through the midpoint of the line segment. In this case, point M is the midpoint of segment AB.

3.3.3. Connecting Angle and Perpendicular Bisectors

Angle bisectors and perpendicular bisectors often intersect at a common point. This intersection point has unique properties and is known as the incenter of a triangle. In the context of a triangle, the incenter is equidistant from all three sides, as it is the point where angle bisectors meet. Additionally, it is the center of the incircle, a circle that can be inscribed within the triangle (Chang, 2002) (Figure 3.5).

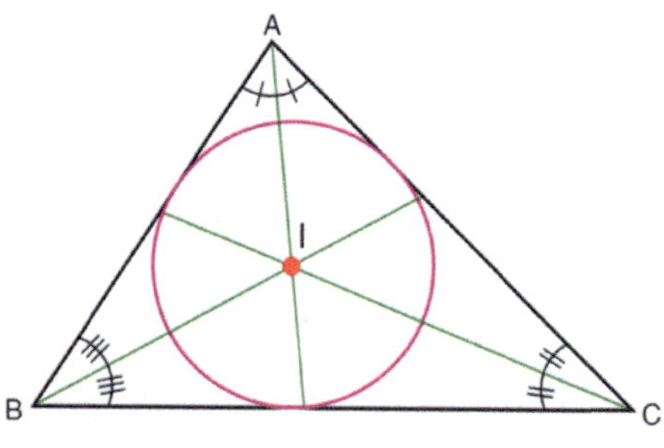

Figure 3.5. Illustration of the intersection of angle and perpendicular bisector.

Source: Math Monks, Creative Commons License.

In Figure 3.5, I is the incenter of triangle ABC. Lines AI, BI, and CI are angle bisectors, and each is a perpendicular bisector of the opposite side. The circle with center I is the incircle of the triangle (Ishikawa, 2019).

- Ø **Example: Constructing the Perpendicular Bisector of a Line Segment:**

Let's say we have a line segment AB, and we want to construct its perpendicular bisector:

1. Locate the midpoint M of segment AB. To find the midpoint, use the midpoint formula:

   $$M = \left(\frac{x_1 + x_2}{2}, \frac{y_1 + y_2}{2}\right)$$

2. Draw a straight-line CD that passes through point M perpendicular to AB. Use a protractor to ensure that the angle formed between CD and AB is 90°.

The line CD is the perpendicular bisector of segment AB, dividing it into two equal parts with M as the midpoint. In summary, angle and perpendicular bisectors are fundamental concepts in geometry, enabling us to divide and connect geometric elements accurately. They have practical applications in construction, navigation, and various mathematical proofs (Gluck, 1991).

3.3.4. Use of Midpoints and Bisectors

Understanding midpoints and bisectors in geometry is not only essential for theoretical knowledge but also for practical applications. These concepts find relevance in various scenarios, from construction to solving complex geometric problems. This section will explore practical examples that illustrate using midpoints and bisectors.

3.3.4.1. Finding the Center of a Room

Imagine you are an architect tasked with designing a circular room. To find the center point for the room's layout, you can use the concept of a perpendicular bisector.

1. **Scenario:** You have the room's diameter, the longest distance between two points on the room's boundary.
2. Procedure:

i. Locate two opposite points on the room's boundary, which define the diameter. Let's call these points A and B.

ii. Use a measuring tape or any measuring tool to find the midpoint of the line segment AB. This midpoint is the center of the circular room.

3. **Mathematical Aspect:** If AB is the diameter of the room, and M is the midpoint, the coordinates of M can be found using the midpoint formula:

$$M = \left(\frac{x_A + x_B}{2}, \frac{y_A + y_B}{2} \right)$$

It ensures that point M lies exactly at the center of the room, making it easier to design the circular layout.

3.3.4.2. Creating Symmetrical Patterns

Symmetry is a fundamental concept in art and design. Artists often use midpoints and bisectors to create symmetric patterns and designs (Rubino et al., 2020).

1. **Scenario:** An artist wants to create a symmetrical design on a canvas.
2. **Procedure:**
   i. Start by drawing the central element or motif of the design.
   ii. Use a straightedge to draw a line, which serves as the axis of symmetry (a perpendicular bisector) passing through the center of the central motif.
   iii. Create a mirror image of the motif on the other side of the line. It is done by ensuring that corresponding points on both sides of the line are equidistant from it, ensuring perfect symmetry (Drury et al., 1996).
3. **Mathematical Aspect:** In this case, the perpendicular bisector serves as the line of reflection, guaranteeing that the distances from the central motif to the line are equal. It ensures that the final design is symmetric.

3.3.4.3. Designing Bridges with Equal Weight Distribution

In civil engineering, especially in bridge design, the concept of midpoints and bisectors is crucial for ensuring the equal distribution of weight and stability (Stucke et al., 2012).

1. **Scenario:** You are a civil engineer tasked with designing a bridge that spans a river.
2. **Procedure:**
   i. Divide the bridge into segments by regularly placing supports (columns or piers). These supports act as midpoints.
   ii. Ensure each support is placed equidistant from its neighboring supports, creating equal-length segments.

iii. Doing this guarantees that the bridge's weight and the load are distributed evenly, preventing structural issues like sagging or collapsing.

3. **Mathematical Aspect:** The concept of midpoints ensures that the bridge segments are of equal length. It can be mathematically verified by measuring the distances between supports and confirming that they are the same.

3.3.4.4. Finding the Geographic Center of a Region

In geography and navigation, midpoints, and bisectors are used to determine the geographic center of a region, which can be useful for various purposes, such as locating a central point for a city (Maneiro Dios & Amatria Jiménez, 2018).

1. **Scenario:** You want to find the geographic center of a large city or region.
2. **Procedure:**

i. Select at least three locations within the city or region (landmarks, coordinates, or known points).

ii. Use a map or GPS coordinates to calculate the midpoint of each pair of selected locations. It involves finding the average latitude and longitude for each pair.

iii. The point where the bisectors of these midpoints intersect is considered the geographic center of the city or region.

3. **Mathematical Aspect:** Calculating the geographic center involves finding the midpoints between known geographic coordinates and ensuring that the bisectors of these midpoints intersect at a single point, representing the city's center.

3.3.4.5. Balancing Rotating Machinery

In engineering, especially in rotating machinery like turbines or propellers, ensuring balance is crucial for efficiency and safety (Costagliola et al., 2014).

1. **Scenario:** You are an engineer responsible for designing a large industrial fan.
2. **Procedure:**

i. Divide the fan blades into equal sections, each separated by a radial line originating from the center of rotation (the midpoint).

ii. Ensure that the mass distribution in each blade section is balanced, meaning that the distances from the center of rotation to the mass elements on each side of the radial line are equal.

iii. Balancing the fan ensures it rotates smoothly and doesn't cause excessive vibrations or wear on the machinery.

3. **Mathematical Aspect:** The concept of midpoints ensures that the radial lines divide the fan blades evenly, while the equal distribution of mass on each side of these radial lines is crucial for balance.

3.4. COLLINEARITY AND COPLANARITY

These concepts play a crucial role in various geometric proofs, theorems, and practical applications, helping to establish relationships between elements in space or on a plane. Understanding collinearity and coplanarity is essential for solving complex geometric problems and scenarios (Kneissl et al., 2011).

3.4.1. Collinear Points and Non-Collinear Points

This section will study the definitions, characteristics, and implications of collinear and non-collinear points, exploring the mathematical language, equations, and examples that elucidate these concepts.

Remember

Collinear points are those that lie on the same straight line. This concept is particularly useful in various fields, from engineering to computer graphics, where understanding whether points are collinear or not can simplify problem-solving.

3.4.1.1. Collinear Points

Collinear points are three or more points that lie on the same straight line. Drawing a line through any two of these points would pass through all the points in the set. Mathematically, given three points A, B, and C, they are collinear if the following condition is met:

$$AB + BC = AC$$

Here, AB represents the distance between point A and point B, BC represents the distance between point B and point C, and AC represents the distance between point A and point C. If this equation holds, then points A,

B, and C are collinear. Essentially, collinear points are co-planar, and the line passing through them is often called a "line of collinearity" (Figure 3.6).

Figure 3.6. *Four points, A, P, Q, and B, lie on the same line, hence are collinear points.*

Source: Math Doubts, Creative Commons License.

The following are the characteristics of collinear points:

1. **Distance Along the Line:** The distances between adjacent points are additive in a set of collinear points. It means that the sum of the distances between any two consecutive points equals the distance between the first and last points in the set.
2. **Unique Line:** Only one straight line can pass through a set of collinear points. The positions of these points uniquely determine this line.
3. **No Intersecting Lines:** A set of collinear points has no other lines passing through all the points except the primary line of collinearity (Olson, 2006).

3.4.1.2. Non-Collinear Points

In contrast, non-collinear points do not lie on the same straight line. It implies that no straight line can simultaneously pass through all the points. Mathematically, if points A, B, and C are not collinear, the condition $AB + BC \neq AC$ holds (Figure 3.7).

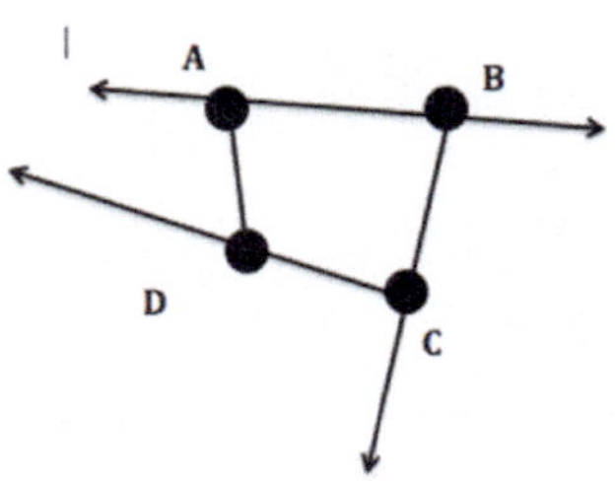

Figure 3.7. Four points A, B, C, and D are non-collinear.

Source: eNotes, Creative Commons License.

The following are the characteristics of non-collinear points:

1. **No Common Line:** Non-collinear points cannot be connected by a straight line. Instead, they are situated in different directions or planes.
2. **Multiple Lines:** Different pairs of non-collinear points may have distinct lines passing through them. In essence, multiple lines can be drawn using these points.

Let's illustrate collinear and non-collinear points through examples (Battersby, 2009).

Ø **Example 1: Collinear Points**

Consider three points, A (2, 3), B(4, 5), and C(6, 7) on the Cartesian plane. To determine if they are collinear, we can calculate the distances between them:

$$AB = \sqrt{(4-2)^2 + (5-3)^2} = \sqrt{4+4} = \sqrt{8}$$

$$BC = \sqrt{(6-4)^2 + (7-5)^2} = \sqrt{4+4} = \sqrt{8}$$

$$AC = \sqrt{(6-2)^2 + (7-3)^2} = \sqrt{16+16} = \sqrt{32} = 4\sqrt{2}$$

Now, we check if the equation $AB + BC = AC$ holds:

$$\sqrt{8} + \sqrt{8} = 4\sqrt{2}$$

Indeed, $4\sqrt{2} = 4\sqrt{2}$

Hence, points A, B, and C, are collinear.

Ø **Example 2: Non-Collinear Points**

Consider three points, D (1, 2), E(3, 4), and F(5, 6) on the Cartesian plane. We can calculate the distances between them as before:

$$DE = \sqrt{(3-1)^2 + (4-2)^2} = \sqrt{4+4} = \sqrt{8}$$

$$EF = \sqrt{(5-3)^2 + (6-4)^2} = \sqrt{4+4} = \sqrt{8}$$

$$DF = \sqrt{(5-1)^2 + (6-2)^2} = \sqrt{16+16} = \sqrt{32} = 4\sqrt{2}$$

Now, let's check the condition $DE + EF \neq DF$:

$$\sqrt{8} + \sqrt{8} \neq 4\sqrt{2}$$

$$2\sqrt{8} \neq 4\sqrt{2}$$

$$4\sqrt{2} \neq 4\sqrt{2}$$

It confirms that points D, E, and F are non-collinear.

Thus, understanding collinearity and non-collinearity is fundamental in various geometric problems, such as determining whether a triangle is degenerate or non-degenerate and establishing the position of points in geometric configurations. These concepts serve as the building blocks for more complex spatial relationships and transformations in geometry (Smith et al., 1987).

3.4.2. Coplanar Points and Non-Coplanar Points

Understanding whether points are coplanar or non-coplanar is essential when working with geometric figures in three-dimensional space.

This section will explore the mathematical definitions, characteristics, and implications of coplanar and non-coplanar points, using equations and examples to illustrate these concepts.

3.4.2.1. Coplanar Points

Coplanar points are a set of points that lie within the same plane. A plane in geometry can be thought of as a flat, two-dimensional surface extending infinitely in all directions. When three or more points lie on the same plane, they are considered coplanar (Robinson, 2003) (Figure 3.8).

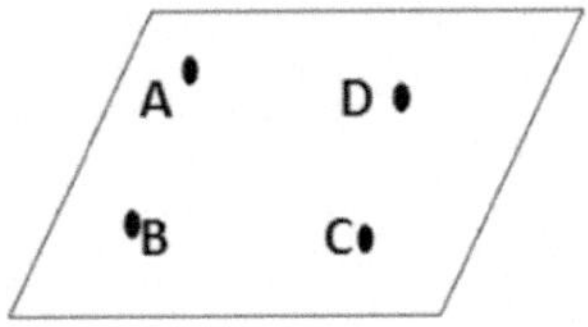

Figure 3.8. Points A, B, C, and D are coplanar.

Source: Elizabeth Often, Creative Commons License.

Mathematically, given four points A, B, C, and D, they are coplanar if the determinant of the matrix formed by their coordinates is zero:

$$\begin{vmatrix} x_A & y_A & z_A & 1 \\ x_B & y_B & z_B & 1 \\ x_C & y_C & z_C & 1 \\ x_D & y_D & z_D & 1 \end{vmatrix} = 0$$

This determinant equation serves as a test for coplanarity. If it equals zero, the points A, B, C, and D, are coplanar; if not, they are non-coplanar (Knox-Robinson & Wyborn, 1997).

The following are the characteristics of coplanar points:

1. **Same Plane:** Coplanar points lie on a common flat surface, or plane, which can be visualized as a sheet of paper extending infinitely.
2. **Determinant Equals Zero:** The mathematical condition for coplanarity is that the determinant of the matrix formed by the coordinates of the points is zero.
3. **Connected Lines:** When three points are coplanar, the lines formed by connecting any pair of these points also lie on the same plane.

3.4.2.2. Non-Coplanar Points

Non-coplanar points are points that do not lie within the same plane. In three-dimensional space, points can be situated so that they cannot all be contained within a single flat surface (Leergaard et al., 2000) (Figure 3.9).

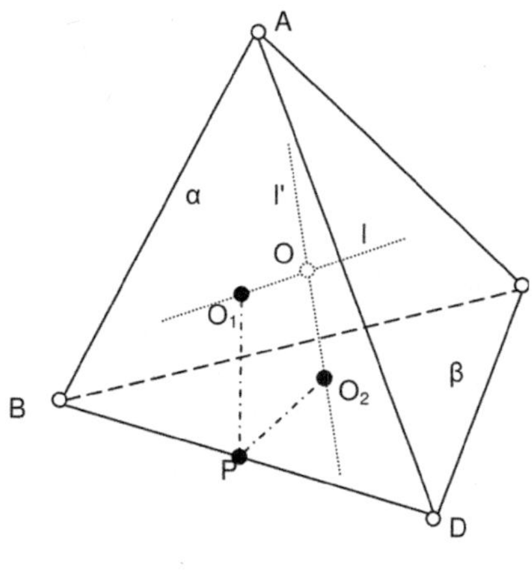

Figure 3.9. Points A, B, C, and D are four non-coplanar points.

Source: Zhenhua Tan, Creative Commons License.

Mathematically, if the determinant of the matrix formed by the coordinates of points A, B, C, and D is non-zero, then they are non-coplanar.

The following are the characteristics of non-coplanar points:

1. **No Common Plane:** Non-coplanar points cannot be contained within a single plane. They are spread out in three-dimensional space.
2. **Determinant Not Equal to Zero:** The mathematical condition for non-coplanarity is that the determinant of the matrix formed by the coordinates of the points is not equal to zero.
3. **Spatial Spread:** Non-coplanar Points can be visualized as being situated in different directions in space. There is no flat surface that contains all of them.

Let's explore some examples to illustrate coplanar and non-coplanar points.

Ø **Example 1: Coplanar Points**

Consider four points A (1, 2, 3), B (3, 4, 5), C (2, 3, 4), and D (4, 5, 6). To determine if they are coplanar, we can use the determinant condition:

$$\begin{vmatrix} 1 & 2 & 3 & 1 \\ 3 & 4 & 5 & 1 \\ 2 & 3 & 4 & 1 \\ 4 & 5 & 6 & 1 \end{vmatrix} = 0$$

This matrix's determinant equals zero, confirming that points A, B, C, and D, are coplanar.

Ø **Example 2: Non-Coplanar Points**

Consider the four points P (1, 2, 3), Q (3, 4, 5), R (2, 3, 4), and S (4, 5, 7). Using the determinant condition:

$$\begin{vmatrix} 1 & 2 & 3 & 1 \\ 3 & 4 & 5 & 1 \\ 2 & 3 & 4 & 1 \\ 4 & 5 & 7 & 1 \end{vmatrix} \neq 0$$

The determinant of this matrix is not equal to zero, indicating that points P, Q, R, and S are non-coplanar.

3.5. PROJECTIONS IN THREE DIMENSIONS

Projections are powerful tools in engineering and architecture, enabling professionals to depict complex 3D structures on flat surfaces such as blueprints and drawings.

There are two main types of projections: orthographic projections and axonometric projections. This section will explore these projection techniques, their mathematical underpinnings, and their practical applications in engineering and architecture (Legleiter & Kyriakidis, 2006).

3.5.1. Orthographic Projections

Orthographic projections, also known as orthogonal projections or multi-view projections, are the most common and widely used type of projection in engineering and architecture.

They provide a systematic way to represent three-dimensional objects using a set of two-dimensional views, each showing the object from a different direction. These views are typically called the front, top, and side views (Figure 3.10).

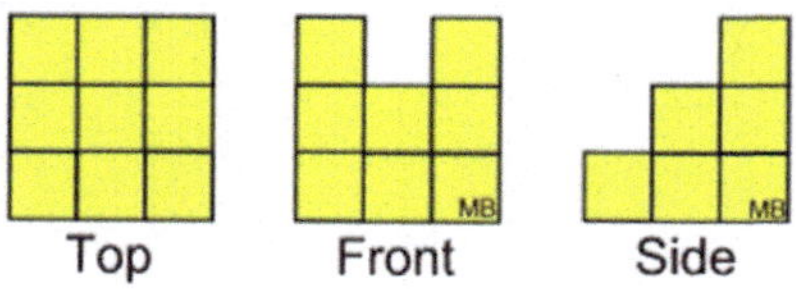

Figure 3.10. Top, front, and side view in orthographic projections.

Source: Math Bits Notebook. Creative Commons License.

Mathematically, orthographic projections involve projecting points from the 3D object onto a 2D plane parallel to one of the coordinate planes (usually the xy, xz, or yz plane). The key idea is that each view captures the object as if illuminated by a set of parallel rays of light emanating from a particular direction (Ma et al., 2018). The front view, for instance, represents the object as seen when looking at it from the front, with the projection plane parallel to the xy-plane. The top view represents the object, with the projection plane parallel to the xz plane. Similarly, the side view represents the object as seen from the side, with the projection plane parallel to the yz-plane. Orthographic projections are a type of technical drawing used in engineering, architecture, and design to represent objects in a two-dimensional space while maintaining accurate proportions and angles. In orthographic projections, objects are typically shown from multiple viewpoints, each projection being a 2D representation of the object from a specific direction (Yang et al., 2021).

Ø **Example: Orthographic Projection of a Cube**

Let's consider a simple scenario where we want to create an orthographic projection of a cube. The cube is positioned in three-dimensional space, and we want to represent its three principal views: front, top, and side (Figure 3.11).

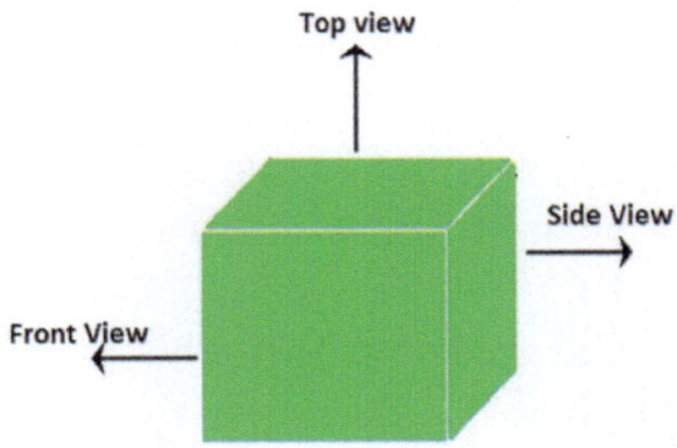

Figure 3.11. Top, front, and side projections of the cube.

Source: Geeks for Geeks, Creative Commons License.

1. **Front View:** It shows the cube as if we are looking at it directly from the front, with one side of the cube hidden.
2. **Top View:** It represents the cube as if we are viewing it directly above, with the top face visible.
3. **Side View:** It shows the cube from the right side, with one of the lateral faces visible.

Orthographic projections involve projecting 3D points onto a 2D plane along specific viewing directions (orthogonal to the plane). The mathematical equations for orthographic projections are relatively simple. Given a 3D point (X, Y, Z) and a viewing direction (e.g., front view), the orthographic projection onto a 2D plane (e.g., the X-Y plane) can be represented as follows:

- **Front View (X-Y Plane):**

$$X\ roj \quad X$$

$$Y\ roj \quad Y$$

- **Top View (X-Z Plane):**

$$X_p roj = X$$

$$Y_p roj = Z$$

- **Side View (Y-Z Plane):**

$$X_proj = Y$$

$$Y_proj = Z$$

In these equations, (X_proj, Y_proj) are the 2D coordinates of the projected point on the respective plane.

For example, if you have a cube with vertices at the corners (0, 0, 0), (1, 0, 0), (1, 1, 0), (0, 1, 0), (0, 0, 1), (1, 0, 1), (1, 1, 1), and (0, 1, 1), you can apply these equations to create orthographic projections for each view by projecting each vertex onto the corresponding 2D plane. The resulting 2D representations accurately capture the shape and size of the cube from different viewpoints (Wu et al., 2019). Orthographic projections are typically accompanied by projection lines, indicating the path from each point on the object to the projection plane. These lines help maintain accuracy and consistency in representing the object's dimensions and angles.

3.5.2. Axonometric Projections

Axonometric projections, sometimes called isometric projections, are another class of projections used to depict three-dimensional objects. Unlike orthographic projections, axonometric projections do not rely on the object's relative position to projection planes. Instead, they involve projecting the object's points along lines that are not necessarily parallel to each other or the coordinate axes.

Did You Know?

Axonometric projections, a type of spatial transformation, can make 3D objects appear more realistic in computer graphics and gaming. They involve skewing the object's dimensions to simulate perspective, making virtual worlds look more immersive.

One of the most common axonometric projections is the isometric projection, which is widely used for technical drawings and illustrations. In an isometric projection, the object is rotated so that all three axes make equal angles with the projection plane. It results in a drawing where all three dimensions (length, width, and height) are represented at the same scale, creating a visually appealing and easily comprehensible representation (Mikolajczyk et al., 2003).

Mathematically, isometric projections are achieved by applying a set of transformation equations to each point on the 3D object to determine its corresponding position in the 2D isometric view. These equations maintain the relative distances and angles between points, ensuring accurate representation.

Ø **Example: Isometric Projection of a Cube**

Let's consider a cube we want to represent using an isometric projection. In an isometric projection, all three axes (X, Y, and Z) are equally foreshortened, which means that all angles between the axes are 120° (Figure 3.12).

To create an isometric projection, we must transform the 3D coordinates of the cube's vertices into 2D coordinates on the isometric plane. Transformation typically involves scaling and rotating the coordinates. Here are the equations for converting 3D coordinates (X, Y, Z) into isometric coordinates $(X_i so, Y_i so)$:

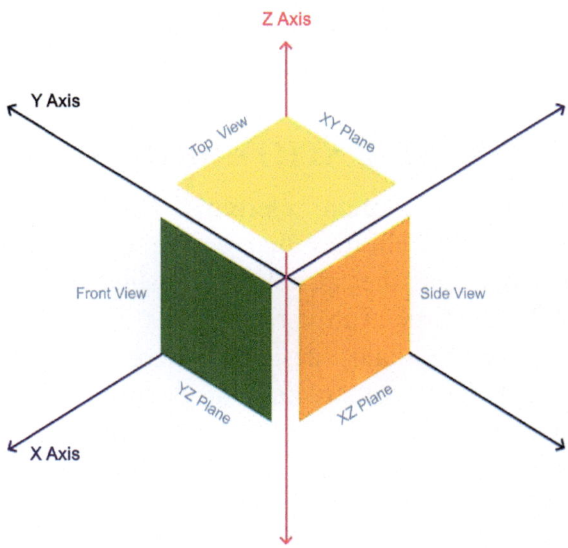

Figure 3.12. Isometric illustration of the cube.

Source: Nitish Khagwal, Creative Commons License.

- **Scale the Coordinates:**

$$X_i so = X \times cos(30) - Y \times sin(30)$$

$$Y_i so = X \times sin(30) + Y \times cos(30)$$

- **Apply Depth (Z) Factor:**

$$Z_i so = Z$$

In these equations, $(X_i so, Y_i so)$ are the isometric coordinates, and (X, Y, Z) are the 3D coordinates of a point on the cube (Erwig & Schneider, 2003).

For example, if you have a cube with vertices at (0, 0, 0), (1, 0, 0), (1, 1, 0), (0, 1, 0), (0, 0, 1), (1, 0, 1), (1, 1, 1), and (0, 1, 1), you can apply these equations to transform the 3D coordinates into isometric coordinates, resulting in a 2D representation of the cube that accurately maintains its proportions in all three dimensions.

This isometric projection provides a simplified and easily understandable representation of the cube, making it a valuable tool for technical and architectural drawings where precise measurements and proportions need to be conveyed without distortion (Jund et al., 2018).

3.6. SPATIAL VISUALIZATION SKILLS

Spatial visualization skills, often referred to as spatial reasoning or intelligence, are cognitive abilities that enable individuals to mentally manipulate and understand the spatial relationships between objects or shapes. These skills are crucial in various fields, including mathematics, engineering, architecture, and design, where comprehending and manipulating three-dimensional structures is essential. In this section, we will study the development of spatial visualization skills, their significance in problem-solving and design, and practical examples of their application (Hollebrands, 2003).

3.6.1. Developing Spatial Visualization Skills

Spatial visualization skills are not innate; they can be developed and refined through practice and training. Individuals with strong spatial visualization skills can mentally rotate and manipulate objects, predict how objects will appear from different angles, and understand complex spatial relationships. Here are some strategies to enhance these skills:

1. **Puzzles and Games:** Engaging in jigsaw puzzles, Rubik's cubes, and spatial reasoning games can help develop spatial visualization skills. These activities require individuals to mentally manipulate pieces to form complete structures (Ambler & Popplestone, 1975).

i. **Example:** You have a jigsaw puzzle with pieces of various shapes, and you want to determine if two pieces fit together.

ii. **Solution:** Check the congruence of edge lengths and angles. For instance, if Piece A has an edge with a length of 'a' and an angle of 'θ,' and Piece B has a corresponding edge with the same length 'a' and angle 'θ,' they will likely fit together.

2. **Architectural Modeling:** Constructing physical models of buildings or structures provides hands-on experience with spatial relationships. It can be particularly valuable for aspiring architects and designers (Lu & Milios, 1997).

i. **Example:** You are creating a linear perspective drawing of a building to represent it on a 2D surface.

ii. **Solution:** Use the linear perspective equation to calculate the vanishing point for lines that recede into the distance. For example, the equation for a line in a one-point perspective might be:

$$x = kx',$$

where; x and x' are the coordinates in 3D and 2D space, respectively.

3. **Geometry and Trigonometry:** Studying geometry and trigonometry, which deal with angles, shapes, and spatial relationships, can improve spatial reasoning. Concepts like congruence, similarity, and transformations are particularly relevant.

i. **Example:** You're solving a geometry problem involving the similarity of two triangles.

ii. **Solution:** Use the similarity ratio to compare the corresponding sides of similar triangles. If triangle ABC resembles triangle DEF, you can write equations like:

AB/DE = AC/DF = BC/EF

3.6.2. Practical Examples of Application

Let's consider practical examples of how spatial visualization skills are applied:

i. **Architectural Design:** Architects use spatial skills to design buildings, create floor plans, and visualize how structures will look and function.

Ø **Example: Golden Ratio (Φ):**

$$\Phi = \left(1 + \sqrt{5}\right) / 2$$

Architects use this ratio to create aesthetically pleasing designs and proportions in buildings.

ii. **Engineering:** Engineers employ spatial reasoning to design machinery, bridges, and electronic circuits. They need to ensure that all components fit together seamlessly.

Ø **Example: Vector Addition**

$$F_{total} = F1 + F2 + F3 + \ldots$$

Engineers use vector addition to calculate the total force acting on a structure by summing individual forces in different directions.

iii. **Product Design:** Industrial and product designers use spatial skills to create prototypes and visualize how products will appear from different angles (Landau et al., 1981).

iv. **Mathematical Problem Solving:** Mathematicians use spatial reasoning to solve geometry problems, understand complex equations, and visualize functions in multivariable calculus.

v. **Medical Imaging:** Radiologists and medical practitioners use spatial skills to interpret 3D medical images, such as CT scans and MRIs.

vi. **Mechanical Repairs:** Mechanics and technicians visualize the inner workings of machinery to diagnose and repair issues effectively (Mark & Egenhofer, 1994).

Hence, spatial visualization skills are essential cognitive abilities that can be developed and have far-reaching applications in various fields. They enable individuals to understand and manipulate complex 3D structures, making them invaluable in problem-solving, design, and everyday life. Whether you are an architect, engineer, mathematician, or someone looking to enhance your problem-solving abilities, cultivating spatial intelligence is valuable.

3.6.3. Spatial Geometry in Computer Graphics and Gaming

Spatial geometry forms the foundation of computer graphics and gaming, where realistic 3D environments and objects are created and manipulated. Here are some key applications:

1. **3D Modeling:** Artists and developers use spatial geometry to create 3D models of characters, objects, and environments in video games, animations, and simulations (Xia et al., 2020).

Ø **Example: Parametric Equations for 3D Objects**

Artists and developers can use parametric equations to describe 3D objects mathematically. For example, the equation for a sphere with radius R centered at the origin is:

$$x = R \times sin(q) \times cos(j)$$

$$y = R \times sin(q) \times sin(j)$$

$$z = R \times cos(q)$$

2. **Rendering:** These engines use spatial geometry to calculate how light interacts with 3D objects, creating realistic shading, shadows, and reflections.

Ø **Example: Phong Reflection Model for Specular Reflection**

$$Specula\ Intensity = Ligh\ Intensity \times (cos(q))^n$$

where; θ is the angle between the light source and the surface normal; and n is the shininess coefficient.

3. **Collision Detection:** In gaming, spatial geometry is essential for detecting collisions between objects, enabling realistic physics simulations and interactions.
4. **Virtual Reality (VR) and Augmented Reality (AR):** These applications use spatial geometry to create immersive experiences. Spatial tracking technology enables users to interact with digital objects in the real world (Tjaden et al., 2016).
5. **Physics Simulations:** Spatial geometry is used in physics simulations for games and scientific research. It allows for the accurate modeling of physical interactions in virtual environments.

6. **Level Design:** Game-level designers use spatial geometry to create game worlds, defining the layout of levels and the placement of objects and obstacles.
7. Camera Systems: Spatial geometry calculates camera positions and views in 3D environments, allowing for dynamic camera movements and perspectives.

Ø **Example: Perspective Projection**

Calculating 2D screen coordinates (x, y) from 3D world coordinates (X, Y, Z) using a perspective projection matrix.

3.6.4. Spatial Relationships in Geographic Information Systems (GIS)

GIS is a field that relies heavily on spatial geometry to analyze and visualize geographical data. It involves using maps, satellite imagery, and spatial data to understand and solve real-world problems. Here are some applications:

1. **Mapping and Cartography:** Spatial geometry is fundamental to creating maps and charts accurately representing geographical features and locations (Wakahara & Yamashita, 2019).

Ø **Example: Mercator Projection**

i. Latitude (φ) to Y-Coordinate:

$$Y = R \times ln\left(tan\left(p / 4 + j \ / 2\right)\right)$$

ii. Longitude (λ) to X-Coordinate:

$$X = R \times l$$

2. **Environmental Analysis:** GIS analyzes environmental data, such as land use, terrain, and vegetation, for urban planning, conservation, and disaster management.

Ø **Example: Digital Elevation Model (DEM) Interpolation (Inverse Distance Weighting)**

$$Z = \Sigma\left(Zi / di^{p}\right) / \Sigma\left(1 / di^{p}\right)$$

where; Z is the estimated elevation; Zi is the elevation at each sampled point; di is the distance between the estimated point and sampled points; and p is the power parameter.

3. **Transportation Planning:** Spatial geometry helps plan transportation networks, optimize routes, and analyze traffic patterns for efficient road and public transportation systems.

Ø **Example: Shortest Path (Dijkstra's Algorithm)**

Distance from source to node v:

$$d[v] = min(d[v], d[u] + weight(u, v))$$

where; d[v] is the shortest distance from the source to node v; u is the current node; and weight (u, v) is the cost of traveling from u to v.

4. **Epidemiology:** Spatial analysis is crucial in epidemiology to track the spread of diseases, identify hot spots, and plan healthcare resources.

Ø **Example: Spatial Interpolation (Kriging)**

$$Z(x) = m + \Sigma l\ i \times j\ i(x)$$

where; Z(x) is the estimated disease prevalence at location x; μ is the global mean; λi are Kriging weights; and φi(x) are spatial basis functions.

5. **Natural Resource Management:** GIS manages natural resources like forests, water bodies, and minerals. It aids in monitoring resource extraction and conservation efforts (Takagi et al., 1989).

Ø **Example: NDVI (Normalized Difference Vegetation Index)**

$$NDVI = (NIR - Red) / (NIR + Red)$$

where; NIR is near-infrared reflectance; and red is red reflectance.

6. **Real Estate and Property Management:** Spatial geometry plays a role in real estate for property assessment, land valuation, and urban development.

Ø **Example: Property Valuation (Regression Model)**

$$Y = b0 + b1 \times X1 + b2 \times X2 + \ldots + e$$

where; Y is the property value; X1, X2, are property attributes (e.g., size, location); β0, β1, β2, are regression coefficients, and ε is the error term.

Activity 3.1

Imagine you're designing a 3D video game environment. You have a building with several windows at different heights and angles. How would you use orthographic projections to accurately represent the placement of these windows in the game world?

3.7. SUMMARY

This chapter provides essential insights into spatial relationships and transformations. It starts by explaining how to measure distances in two and three dimensions, which has practical applications in various real-life situations. The concept of midpoint and bisectors is also covered, offering a simple way to find the midpoint between two points and understand angle bisectors. Collinearity and coplanarity are discussed to distinguish points on the same line or plane. The chapter introduces three-dimensional projections, such as orthographic and axonometric projections, which are valuable in computer graphics. Finally, it emphasizes developing spatial visualization skills and their applications, including in computer graphics and geographic information systems.

REFERENCES

1. Ahmed, N., & Miller, H. J., (2007). Time–space transformations of geographic space for exploring, analyzing, and visualizing transportation systems. *Journal of Transport Geography*, *15*(1), 2–17.
2. Ambler, A. P., & Popplestone, R. J., (1975). Inferring the positions of bodies from specified spatial relationships. *Artificial Intelligence*, *6*(2), 157–174.
3. Aminova, A. V., (1995). Projective transformations and symmetries of differential equation. *Sbornik: Mathematics*, *186*(12), 1711.
4. Aslaksen, H., (1991). Restricted homogeneous coordinates for the Cayley projective plane. *Geometriae Dedicata*, *40*(2), 245–250.
5. Balakrishnan, G., Zhao, A., Sabuncu, M. R., Guttag, J., & Dalca, A. V., (2018). An unsupervised learning model for deformable medical image registration. In: *Proceedings of the IEEE Conference on Computer Vision and Pattern Recognition* (Vol. 3, pp. 9252–9260).
6. Battersby, S. E., (2009). The effect of global-scale map-projection knowledge on perceived land area. *Cartographica: The International Journal for Geographic Information and Geovisualization*, *44*(1), 33–44.
7. Belghit, H., Zenati-Henda, N., Bellabi, A., Benbelkacem, S., & Belhocine, M., (2012). Tracking color marker using projective transformation for augmented reality application. In: *2012 International Conference on Multimedia Computing and Systems* (Vol. 2, pp. 372–377). IEEE.
8. Bostanci, E., Kanwal, N., & Clark, A. F., (2012). Feature coverage for better homography estimation: An application to image stitching. In: *2012 19th International Conference on Systems, Signals, and Image Processing (IWSSIP)* (Vol. 2, pp. 448–451). IEEE.
9. Bryant, D. J., (1998). Human spatial concepts reflect regularities of the physical world and human body. *Representation and Processing of Spatial Expressions*, *10*(5), 215–230.
10. Cantù, C., (2010). Exploring the role of spatial relationships to transform knowledge in a business idea—Beyond a geographic proximity. *Industrial Marketing Management*, *39*(6), 887–897.
11. Cao, Y., Cervone, G., Barkley, Z., Lauvaux, T., Deng, A., & Taylor, A., (2017). Analysis of errors introduced by geographic coordinate systems on weather numeric prediction modeling. *Geoscientific Model Development*, *10*(9), 3425–3440.

12. Capel, D., & Zisserman, A., (2003). Computer vision applied to super resolution. *IEEE Signal Processing Magazine*, *20*(3), 75–86.
13. Chang, M., (2002). A 3D slope stability analysis method assuming parallel lines of intersection and differential straining of block contacts. *Canadian Geotechnical Journal*, *39*(4), 799–811.
14. Chernyavsky, D., & Sorokin, D., (2019). Three-dimensional (higher-spin) gravities with extended Schrödinger and l-conformal Galilean symmetries. *Journal of High Energy Physics*, *2019*(7), 1–29.
15. Chong, C. F., Wang, Y., Ng, B., Luo, W., & Yang, X., (2023). Image projective transformation rectification with synthetic data for smartphone-captured chest X-ray photos classification. *Computers in Biology and Medicine*, *987*(34), 107277.
16. Clark, P. J., & Evans, F. C., (1954). Distance to nearest neighbor as a measure of spatial relationships in populations. *Ecology*, *35*(4), 445–453.
17. Costagliola, G., De Rosa, M., & Fuccella, V., (2014). Recognition and autocompletion of partially drawn symbols by using polar histograms as spatial relation descriptors. *Computers & Graphics*, *39*, 101–116.
18. David, L. T., (2012). Training effects on mental rotation, spatial orientation and spatial visualization depending on the initial level of spatial abilities. *Procedia-Social and Behavioral Sciences*, *33*, 328–332.
19. Demetriou, O., (2006). Streets not named: Discursive dead ends and the politics of orientation in intercommunal spatial relations in Northern Greece. *Cultural Anthropology*, *21*(2), 295–321.
20. Denardo, C., Dougherty, M., Hastie, G., Leaper, R., Wilson, B., & Thompson, P. M., (2001). A new technique to measure spatial relationships within groups of free-ranging coastal cetaceans. *Journal of Applied Ecology*, *38*(4), 888–895.
21. Drury, H. A., Van Essen, D. C., Anderson, C. H., Lee, C. W., Coogan, T. A., & Lewis, J. W., (1996). Computerized mappings of the cerebral cortex: A multiresolution flattening method and a surface-based coordinate system. *Journal of Cognitive Neuroscience*, *8*(1), 1–28.
22. El Mallahi, M., El Mekkaoui, J., Zouhri, A., Amakdouf, H., & Qjidaa, H., (2018). Rotation scaling and translation invariants of 3D radial shifted Legendre moments. *International Journal of Automation and Computing*, *15*, 169–180.

23. Erwig, M., & Schneider, M., (2003). A visual language for the evolution of spatial relationships and its translation into a spatio-temporal calculus. *Journal of Visual Languages & Computing, 14*(2), 181–211.
24. Fossen, H., & Tikoff, B., (1993). The deformation matrix for simultaneous simple shearing, pure shearing and volume change, and its application to transpression-transtension tectonics. *Journal of Structural Geology, 15*(3–5), 413–422.
25. Gleicher, M., (1997). Projective registration with difference decomposition. In: *Proceedings of IEEE Computer Society Conference on Computer Vision and Pattern Recognition* (Vol. 2, pp. 331–337). IEEE.
26. Gleiss, A. C., Norman, B., & Wilson, R. P., (2011). Moved by that sinking feeling: variable diving geometry underlies movement strategies in whale sharks. *Functional Ecology, 25*(3), 595–607.
27. Gluck, M., (1991). Making sense of human wayfinding: review of cognitive and linguistic knowledge for personal navigation with a new research direction. *Cognitive and Linguistic Aspects of Geographic Space, 10*, 117–135.
28. Hanrahan, P., (1989). Ray-triangle and ray-quadrirateral intersections in homogeneous coordinates. *Coordinates, 2*(1), 2.
29. Harris, J., Hirsh-Pasek, K., & Newcombe, N. S., (2013). Understanding spatial transformations: Similarities and differences between mental rotation and mental folding. *Cognitive Processing, 14*, 105–115.
30. Hartley, R. I., & Mundy, J. L., (1993). Relationship between photogrammmetry and computer vision. *Integrating Photogrammetric Techniques with Scene Analysis and Machine Vision, 1944*, 92–105.
31. Hegarty, M., & Kozhevnikov, M., (1999). Types of visual–spatial representations and mathematical problem solving. *Journal of Educational Psychology, 91*(4), 684.
32. Heuel, S., & Forstner, W., (2001). Matching, reconstructing, and grouping 3D lines from multiple views using uncertain projective geometry. In: *Proceedings of the 2001 IEEE Computer Society Conference on Computer Vision and Pattern Recognition. CVPR 2001* (Vol. 2, pp. 56–59). IEEE.
33. Hill, D. L., & Batchelor, P., (2001). Registration methodology: Concepts and algorithms. *Medical Image Registration, 40*, 39–70.

34. Hollebrands, K. F., (2003). High school students' understandings of geometric transformations in the context of a technological environment. *The Journal of Mathematical Behavior*, *22*(1), 55–72.
35. Ishikawa, T., (2019). Satellite navigation and geospatial awareness: Long-term effects of using navigation tools on wayfinding and spatial orientation. *The Professional Geographer*, *71*(2), 197–209.
36. Jund, P., Eitel, A., Abdo, N., & Burgard, W., (2018). Optimization beyond the convolution: Generalizing spatial relations with end-to-end metric learning. In: *2018 IEEE International Conference on Robotics and Automation (ICRA)* (Vol. 2, pp. 4510–4516). IEEE.
37. Kneissl, T., van Gasselt, S., & Neukum, G., (2011). Map-projection-independent crater size-frequency determination in GIS environments—New software tool for ArcGIS. *Planetary and Space Science*, *59*(11–12), 1243–1254.
38. Knox-Robinson, C. M., & Wyborn, L. A. I., (1997). Towards a holistic exploration strategy: Using geographic information systems as a tool to enhance exploration. *Australian Journal of Earth Sciences*, *44*(4), 453–463.
39. Koya, C. F., (2017). The digital vā: Negotiating socio-spatial relations in cyberspace, place, and time. In: *The Relational Self: Decolonizing Personhood in the Pacific* (Vol. 2, pp. 61–78). USP and The Pacific Theological College.
40. Kučerová, S. R., Bláha, J. D., & Kučera, Z., (2015). Transformations of spatial relationships in elementary education: A case study of changes in two Czech rural areas since the second half of the 20th century. *Moravian Geographical Reports*, *23*(1), 34–44.
41. Landau, B., Gleitman, H., & Spelke, E., (1981). Spatial knowledge and geometric representation in a child blind from birth. *Science*, *213*(4513), 1275–1278.
42. Leergaard, T. B., Lyngstad, K. A., Thompson, J. H., Taeymans, S., Vos, B. P., De Schutter, E., & Bjaalie, J. G., (2000). Rat somatosensory cerebropontocerebellar pathways: Spatial relationships of the somatotopic map of the primary somatosensory cortex are preserved in a three-dimensional clustered pontine map. *Journal of Comparative Neurology*, *422*(2), 246–266.
43. Legleiter, C. J., & Kyriakidis, P. C., (2006). Forward and inverse transformations between Cartesian and channel-fitted coordinate systems for meandering rivers. *Mathematical Geology*, *38*, 927–958.

44. Li, H., Hestenes, D., & Rockwood, A., (2001). Generalized homogeneous coordinates for computational geometry. *Geometric Computing with Clifford Algebras: Theoretical Foundations and Applications in Computer Vision and Robotics*, *8*(45), 27–59.
45. Liao, K. H., (2017). The abilities of understanding spatial relations, spatial orientation, and spatial visualization affect 3D product design performance: Using carton box design as an example. *International Journal of Technology and Design Education*, *27*, 131–147.
46. Liu, Y., Guo, Q., & Kelly, M., (2008). A framework of region-based spatial relations for non-overlapping features and its application in object-based image analysis. *ISPRS Journal of Photogrammetry and Remote Sensing*, *63*(4), 461–475.
47. Loewner, C., & Nirenberg, L., (1974). Partial differential equations invariant under conformal or projective transformations. In: *Contributions to Analysis* (Vol. 2, pp. 245–272). Academic Press.
48. Lu, F., & Milios, E., (1997). Globally consistent range scan alignment for environment mapping. *Autonomous Robots*, *4*, 333–349.
49. Ma, W., Xiong, H., Dai, X., Zheng, X., & Zhou, Y., (2018). An indoor scene recognition-based 3D registration mechanism for real-time AR-GIS visualization in mobile applications. *ISPRS International Journal of Geo-Information*, *7*(3), 112.
50. Maneiro Dios, R., & Amatria Jiménez, M., (2018). Polar coordinate analysis of relationships with teammates, areas of the pitch, and dynamic play in soccer: a study of Xabi Alonso. *Frontiers in Psychology*, *9*, 389.
51. Mann, S., (1992). Wavelets and "Chirplets": Time–frequency "perspectives" with applications. In: *Advances in Machine Vision: Strategies and Applications* (Vol. 2, pp. 99–128).
52. Mardia, K. V., Goodall, C., & Walder, A., (1996). Distributions of projective invariants and model-based machine vision. *Advances in Applied Probability*, *28*(3), 641–661.
53. Marinakis, A., Mania, K., & Antoniadis, A., (2021). Augmented reality for cad-cam training featuring 3D interactive geometric transformations. *Computer-Aided Design and Applications*, *18*, 561–570.
54. Marino, R. A., Rodgers, C. K., Levy, R., & Munoz, D. P., (2008). Spatial relationships of visuomotor transformations in the superior colliculus map. *Journal of Neurophysiology*, *100*(5), 2564–2576.

55. Mark, D. M., & Egenhofer, M. J., (1994). Modeling spatial relations between lines and regions: Combining formal mathematical models and human subjects testing. *Cartography and Geographic Information Systems*, *21*(4), 195–212.
56. Mikolajczyk, K., Zisserman, A., & Schmid, C., (2003). Shape recognition with edge-based features. In: *British Machine Vision Conference (BMVC'03)* (Vol. 2, pp. 779–788). The British Machine Vision Association.
57. Mohr, R., (1999). Projective geometry and computer vision. *Handbook of Pattern Recognition and Computer Vision*, *87*, 313–337.
58. Nabil, M., Shepherd, J., & Ngu, A. H., (1995). 2D projection interval relationships: A symbolic representation of spatial relationships. In: *International Symposium on Spatial Databases* (Vol. 2, pp. 292–309). Berlin, Heidelberg: Springer Berlin Heidelberg.
59. Nakano, K., Okutomi, M., & Hasegawa, Y., (2002). Camera calibration with precise extraction of feature points using projective transformation. In: *Proceedings 2002 IEEE International Conference on Robotics and Automation (Cat. No. 02CH37292)* (Vol. 3, pp. 2532–2538). IEEE.
60. Ng, O. L., & Chan, T., (2019). Learning as making: Using 3D computer-aided design to enhance the learning of shape and space in STEM-integrated ways. *British Journal of Educational Technology*, *50*(1), 294–308.
61. Nicholson, J. R., & Seddon, G. M., (1977). The understanding of pictorial spatial relationships by Nigerian secondary school students. *Journal of Cross-Cultural Psychology*, *8*(4), 381–400.
62. Nie, L., Lin, C., Liao, K., Liu, M., & Zhao, Y., (2020). A view-free image stitching network based on global homography. *Journal of Visual Communication and Image Representation*, *73*, 102950.
63. Nielsen, H. P., (1995). Line clipping using semi-homogeneous coordinates. In: *Computer Graphics Forum* (Vol. 14, pp. 3–16). Edinburgh, UK: Blackwell Science Ltd.
64. Olson, J. M., (2006). Map projections and the visual detective: How to tell if a map is equal-area, conformal, or neither. *Journal of Geography*, *105*(1), 13–32.
65. Pairet, È., Hernández, J. D., Carreras, M., Petillot, Y., & Lahijanian, M., (2021). Online mapping and motion planning under uncertainty

for safe navigation in unknown environments. *IEEE Transactions on Automation Science and Engineering, 19*(4), 3356–3378.

66. Panorkou, N., & Maloney, A., (2015). Elementary students' construction of geometric transformation reasoning in a dynamic animation environment. *Constructivist Foundations, 10*(3), 338–347.
67. Papadias, D., & Theodoridis, Y., (1997). Spatial relations, minimum bounding rectangles, and spatial data structures. *International Journal of Geographical Information Science, 11*(2), 111–138.
68. Pasko, A., Adzhiev, V., Sourin, A., & Savchenko, V., (1995). Function representation in geometric modeling: concepts, implementation, and applications. *The Visual Computer, 11*, 429–446.
69. Patro, K., Nuerk, H. C., Cress, U., & Haman, M., (2014). How number-space relationships are assessed before formal schooling: A taxonomy proposal. *Frontiers in Psychology, 5*, 419.
70. Paul, U., Subramanian, A. P., Buddhikot, M. M., & Das, S. R., (2012). Understanding spatial relationships in resource usage in cellular data networks. In: *2012 Proceedings IEEE INFOCOM Workshops* (Vol. 2, pp. 244–249). IEEE.
71. Pea, R. D., (2006). Video-as-data and digital video manipulation techniques for transforming learning sciences research, education, and other cultural practices. In: *The International Handbook of Virtual Learning Environments* (Vol. 2, pp. 1321–1393). Dordrecht: Springer Netherlands.
72. Qiao, J., Zhu, Y., Jia, X., & Huang, L., (2018). Estimating the spatial relationships between soil hydraulic properties and soil physical properties in the critical zone (0–100 m) on the Loess Plateau, China: A state-space modeling approach. *Catena, 160*, 385–393.
73. Riva, G., Baños, R. M., Botella, C., Mantovani, F., & Gaggioli, A., (2016). Transforming experience: The potential of augmented reality and virtual reality for enhancing personal and clinical change. *Frontiers in Psychiatry, 7*, 164.
74. Robinson, V. B., (2003). A perspective on the fundamentals of fuzzy sets and their use in geographic information systems. *Transactions in GIS, 7*(1), 3–30.
75. Rubino, I., Coscia, C., & Curto, R., (2020). Identifying spatial relationships between built heritage resources and short-term rentals before the COVID-19 pandemic: Exploratory perspectives on sustainability issues. *Sustainability, 12*(11), 4533.

76. Schinko, C., Krispel, U., Eggeling, E., & Ullrich, T., (2017). 3D model representations and transformations in the context of computer-aided design: A state-of-the-art overview. In: *Proceedings of the International Conference on Advances in Multimedia* (Vol. 2, pp. 10–15).
77. Schulz, A. J., Williams, D. R., Israel, B. A., & Lempert, L. B., (2002). Racial and spatial relations as fundamental determinants of health in Detroit. *The Milbank Quarterly*, *80*(4), 677–707.
78. Sekhavat, S., Svestka, P., Laumond, J. P., & Overmars, M. H., (1998). Multilevel path planning for nonholonomic robots using semiholonomic subsystems. *The International Journal of Robotics Research*, *17*(8), 840–857.
79. Shelton, B. E., & Hedley, N. R., (2004). Exploring a cognitive basis for learning spatial relationships with augmented reality. *Technology, Instruction, Cognition, and Learning*, *1*(4), 323.
80. Shu, F., Wang, J., Pagani, A., & Stricker, D., (2023). Structure PLP-slam: Efficient sparse mapping and localization using point, line, and plane for monocular, RGB-D, and stereo cameras. In: *2023 IEEE International Conference on Robotics and Automation (ICRA)* (Vol. 2, pp. 2105–2112). IEEE.
81. Skala, V., (2005). A new approach to line and line segment clipping in homogeneous coordinates. *The Visual Computer*, *21*, 905–914.
82. Smith, T. R., Menon, S., Star, J. L., & Estes, J. E., (1987). Requirements and principles for the implementation and construction of large-scale geographic information systems. *International Journal of Geographical Information System*, *1*(1), 13–31.
83. Solari, D., Magro, F., Cappabianca, P., Cavallo, L. M., Samii, A., Esposito, F., & Samii, M., (2007). Anatomical study of the pterygopalatine fossa using an endoscopic endonasal approach: Spatial relations and distances between surgical landmarks. *Journal of Neurosurgery*, *106*(1), 157–163.
84. Song, D. Y., Um, G. M., Lee, H. K., & Cho, D., (2021). End-to-end image stitching network via multi-homography estimation. *IEEE Signal Processing Letters*, *28*, 763–767.
85. Sorrows, M. E., & Hirtle, S. C., (1999). The nature of landmarks for real and electronic spaces. In: *Spatial Information Theory. Cognitive and Computational Foundations of Geographic Information Science: International Conference COSIT'99 Stade, Germany, Proceedings 4* (Vol. 2, pp. 37–50). Springer Berlin Heidelberg.

86. Spring, K. W., (1986). Euler parameters and the use of quaternion algebra in the manipulation of finite rotations: A review. *Mechanism and Machine Theory*, *21*(5), 365–373.

87. Stucke, S., McFarland, D., Goss, L., Fonov, S., McMillan, G. R., Tucker, A., & Davis, B. L., (2012). Spatial relationships between shearing stresses and pressure on the plantar skin surface during gait. *Journal of Biomechanics*, *45*(3), 619–622.

88. Takagi, M., Hiyama, T., Sone, M., & Onoe, M., (1989). Automatic correcting method of geometric distortion in meteorological satellite NOAA image. *Systems and Computers in Japan*, *20*(8), 1–14.

89. Tanaka, N., (1957). Projective connections and projective transformations. *Nagoya Mathematical Journal*, *12*, 1–24.

90. Tjaden, B., Cooper, S. J., Brett, D. J., Kramer, D., & Shearing, P. R., (2016). On the origin and application of the Bruggeman correlation for analyzing transport phenomena in electrochemical systems. *Current Opinion in Chemical Engineering*, *12*, 44–51.

91. Trafton, J. G., Trickett, S. B., Stitzlein, C. A., Saner, L., Schunn, C. D., & Kirschenbaum, S. S., (2006). The relationship between spatial transformations and iconic gestures. *Spatial Cognition and Computation*, *6*(1), 1–29.

92. Uygur, S. Ö., Aydin, C., & Akyilmaz, O., (2022). Retrieval of Euler rotation angles from 3D similarity transformation based on quaternions. *Journal of Spatial Science*, *67*(2), 255–272.

93. Wakahara, T., & Yamashita, Y., (2019). Image alignment using norm conserved GAT correlation. In: *2019 Digital Image Computing: Techniques and Applications (DICTA)* (Vol. 3, pp. 1–6). IEEE.

94. Wang, Y., & Tsui, B. M., (2007). Pinhole SPECT with different data acquisition geometries: Usefulness of unified projection operators in homogeneous coordinates. *IEEE Transactions on Medical Imaging*, *26*(3), 298–308.

95. Wu, Y., Liu, Y., Ahmed, S. H., Peng, J., & Abd El-Latif, A. A., (2019). Dominant data set selection algorithms for electricity consumption time-series data analysis based on affine transformation. *IEEE Internet of Things Journal*, *7*(5), 4347–4360.

96. Xia, C., Yeh, A. G. O., & Zhang, A., (2020). Analyzing spatial relationships between urban land use intensity and urban vitality at street block level: A case study of five Chinese megacities. *Landscape and Urban Planning, 193*, 103669.
97. Yan, B., Xu, N., Zhao, W. B., & Xu, L. P., (2019). A three-dimensional Hough transform-based track-before-detect technique for detecting extended targets in strong clutter backgrounds. *Sensors, 19*(4), 881.
98. Yang, J., Shi, Y., Tong, X., Wang, R., Chen, T., & Ying, X., (2021). Improving knowledge graph embedding using affine transformations of entities corresponding to each relation. In: *Findings of the Association for Computational Linguistics: EMNLP 2021* (Vol. 2, pp. 508–517).
99. Ye, Y., (1990). A class of projective transformations for linear programming. *SIAM Journal on Computing, 19*(3), 457–466.
100. Yiu, P., (2000). The uses of homogeneous barycentric coordinates in plane Euclidean geometry. *International Journal of Mathematical Education in Science and Technology, 31*(4), 569–578.
101. Zhan, W., Hong, S., Sun, Y., & Zhu, C., (2021). The system research and implementation for autorecognition of the ship draft via the UAV. *International Journal of Antennas and Propagation, 2021*, 1–11.
102. Zhu, K., Xu, J., & Wang, X., (2022). The evolution of urban innovation space and its spatial relationships with talents' living demands: evidence from Hangzhou, China. *International Journal of Urban Sciences, 2*, 1–19.
103. Zokai, S., & Wolberg, G., (2005). Image registration using log-polar mappings for recovery of large-scale similarity and projective transformations. *IEEE Transactions on Image Processing, 14*(10), 1422–1434.

CHAPTER 4

Coordinate Geometry

Contents

4.1. INTRODUCTION

Coordinate Geometry, often called Analytical Geometry, is a branch of mathematics that merges algebraic techniques with geometric concepts. Using numerical coordinates provides a powerful framework for exploring geometric shapes, lines, curves, and relationships. This chapter discusses the foundational aspects of coordinate geometry, shedding light on its historical evolution, the Cartesian Plane, various coordinate systems, and essential terminology (Saha et al., 2010). Coordinate Geometry enables us to represent geometric figures and solve geometric problems using algebraic equations and numerical coordinates. In this approach, points in space are described using ordered pairs of numbers, which serve as coordinates. These coordinates provide a systematic and precise way to locate and manipulate geometric objects. The primary aim of coordinate geometry is to facilitate the study of geometric shapes and their attributes by translating them into algebraic expressions. Doing so enables us to analyze geometric relationships, compute distances, and derive equations that describe curves and lines (Yang et al., 2007).

4.1.1. Historical Background

The roots of coordinate geometry can be traced back to ancient Greece, with contributions from mathematicians like Apollonius of Perga, who explored conic sections using geometrical methods. However, the modern development of coordinate geometry is largely attributed to René Descartes, a French mathematician, philosopher, and scientist, in the 17th century. Descartes introduced the Cartesian coordinate system concept in his work "La Géométrie," published in 1637. This revolutionary idea laid the foundation for modern mathematics by providing a systematic method for linking algebraic equations with geometric shapes. It marked a turning point in how mathematics was approached and applied to understand the physical world (Earnest, 2015).

4.1.2. The Cartesian Plane

The Cartesian Plane, named after René Descartes, is the foundation of coordinate geometry. The Cartesian Plane is a two-dimensional plane divided into four quadrants, labeled as the first, second, third, and fourth quadrants, counterclockwise from the upper right. It comprises two perpendicular axes, the horizontal x-axis and the vertical y-axis, intersecting at the origin (0, 0).

The coordinates of any point on the Cartesian Plane are represented as an ordered pair (x, y), where 'x' denotes the distance along the x-axis, and 'y' denotes the distance along the y-axis. These coordinates provide a unique address for each point on the plane, enabling precise location identification axes (Voronina, 2021) (Figure 4.1).

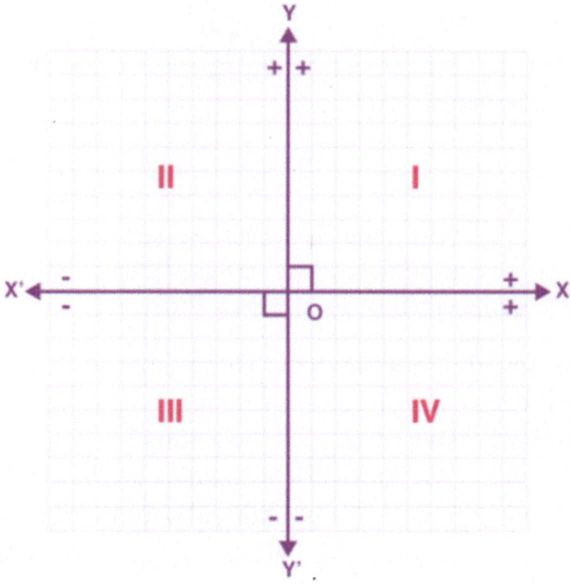

Figure 4.1. Illustration of the Cartesian plane.

Source: BYJU's, Creative Commons License.

4.1.3. Coordinate Systems

While the Cartesian Plane is the most common coordinate system, other coordinate systems exist for specialized applications. Three notable coordinate systems are:

1. **Polar Coordinates:** In this system, points are defined by their distance from the origin (r) and the angle (θ) they make with the positive x-axis. Polar coordinates are particularly useful when circular or radial symmetry is prevalent (Figure 4.2).

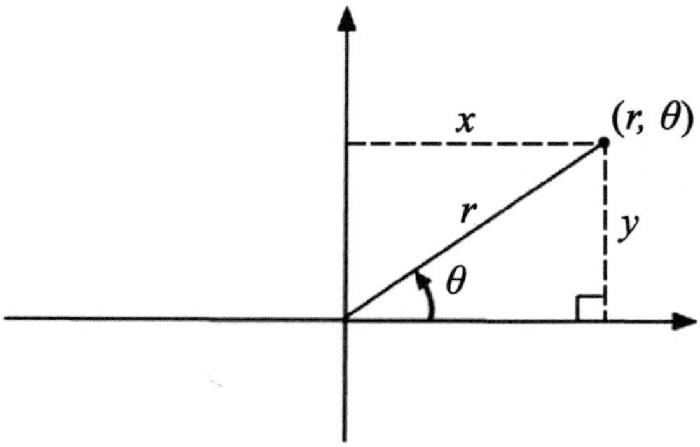

Figure 4.2. Schematic of the polar coordinate system.

Source: BYJU's, Creative Commons License.

2. **Cylindrical Coordinates:** These are an extension of polar coordinates into three dimensions. They introduce a third coordinate, denoted as 'z,' representing the height above or below the xy-plane. Cylindrical coordinates are valuable for problems involving cylindrical or cylindrical-symmetrical objects.
3. **Spherical Coordinates:** These are used in three-dimensional space to represent points using three parameters: radial distance (ρ), polar angle (θ), and azimuthal angle (φ). They are especially effective when dealing with spherical, or spherical-symmetrical objects (Zou et al., 2022) (Figure 4.3).

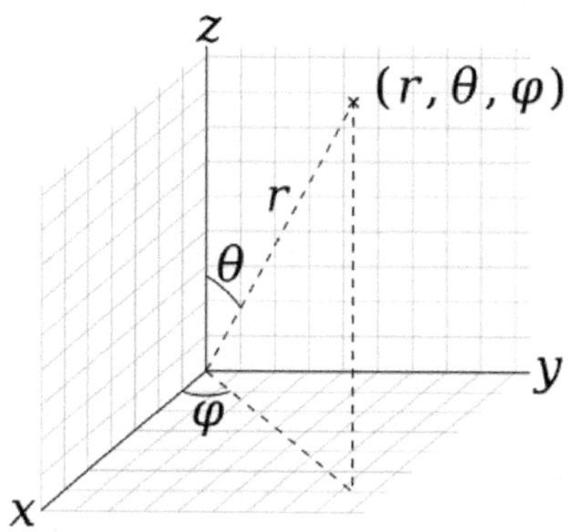

Figure 4.3. Schematic of the spherical coordinate system.

Source: Andeggs, Creative Commons License.

Each coordinate system offers a unique perspective on space and can simplify problem-solving in specific contexts.

4.1.4. Notation and Terminology

Coordinate geometry employs specific notation and terminology to describe points, lines, and geometric objects:

1. **Ordered Pair:** A set of two numbers (x, y) representing the coordinates of a point on the Cartesian Plane.
2. **Origin:** The point (0, 0) where the x-axis and y-axis intersect.
3. **Axes:** The two perpendicular lines on the Cartesian Plane. The horizontal line is the x-axis, and the vertical line is the y-axis.
4. **Quadrants:** The four regions into which the Cartesian Plane is divided.
5. **Gradient (Slope):** A measure of the steepness of a line. The gradient 'm' of a line passing through two points (x_1, y_1) and (x_2, y_2) is given by:

$$m = \frac{y^{2} - y^{1}}{x^{2}, -x^{1}}$$

6. **Intercept:** The point(s) where a line crosses the x-axis or y-axis. The x-intercept is where the line crosses the x-axis (y = 0), and the y-intercept is where it crosses the y-axis (x = 0).

The above introduction sets the stage for an in-depth exploration of coordinate geometry. In the following sections, we will delve into the practical applications of plotting points, deriving equations of lines, calculating distances, and solving geometric problems using the principles of coordinate geometry (Bakuła, 2013).

4.2. PLOTTING POINTS AND LINES ON A PLANE

Plotting points and lines on a plane is a fundamental concept in coordinate geometry. It involves using numerical coordinates to precisely locate points in a two-dimensional space represented by the Cartesian Plane. Each point is denoted by an ordered pair (x, y), where 'x' represents the horizontal distance along the x-axis, and 'y' represents the vertical distance along the y-axis. By plotting these points, geometric shapes and lines can be constructed. Lines, in particular, are determined by two distinct points, and their equations can be derived using the coordinates. Plotting and working with points and lines form the basis for solving geometric problems and exploring spatial relationships in mathematics and beyond.

4.2.1. Plotting Points

In coordinate geometry, plotting points is the fundamental process of locating specific positions within a Cartesian plane. It is essential for understanding spatial relationships, performing geometric analyses, and solving mathematical problems. This section delves into the precise art of plotting points on a Cartesian plane (Timmons et al., 1980).

4.2.1.1. Representation of Points

Points in a Cartesian plane is represented by ordered pairs of numbers, denoted as (x, y). The x-coordinate (abscissa) specifies the horizontal position of the point concerning the origin, while the y-coordinate (ordinate) specifies the vertical position. The intersection of these two coordinates uniquely defines the location of the point within the plane.

For example, consider point A (3, 4). The number 3 represents the horizontal displacement from the origin along the x-axis, indicating that the point is three units to the right of the origin. Simultaneously, the number 4 represents the vertical displacement along the y-axis, indicating that the point is four units above the origin (Thompson, 1967).

4.2.1.2. Plotting Points on a Cartesian Plane

Plotting points on a Cartesian plane involves two main steps: locating the point along the x-axis and then along the y-axis.

1. **Locating the x-Coordinate:** Begin by locating the horizontal position of the point along the x-axis. If the x-coordinate is positive, move to the right of the origin; if it is negative, move to the left. For example, if the x-coordinate is 5, move five units to the right; if it is –2, move two units to the left.
2. **Locating the y-Coordinate:** After establishing the horizontal position, proceed to locate the point along the y-axis. If the y-coordinate is positive, move upward from the horizontal position; if it is negative, move downward. For example, if the y-coordinate is 3, move three units up from the horizontal position, and if it is –1, move one unit down (Niehrs, 2010).

To illustrate this process, let's plot a few points (Figure 4.4):

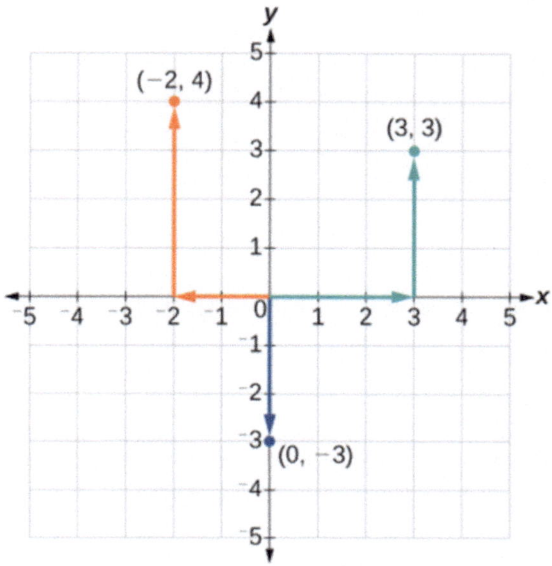

Figure 4.4. Illustration of points on a Cartesian plane.

Source: Lumen Learning, Creative Commons License.

1. **Point (3, 3):** Start at the origin, move three units to the right along the x-axis, then move three units up along the y-axis. You arrive at the point (3, 3).
2. **Point (–2, 4):** Begin at the origin, move two units to the left along the x-axis, and then move four units upward along the y-axis. You locate point C (–2, 4).
3. **Point D (0, –3):** Begin at the origin and move three units below along the y-axis. It represents the point (0, –3).

4.2.2. Equations of Lines

In coordinate geometry, lines play a fundamental role in expressing linear relationships between points and solving various mathematical and real-world problems. Understanding how to represent lines using equations is a crucial skill. In this section, we will delve into the equations of lines, exploring their forms, properties, and applications (Carrasco-Navarro et al., 2022).

A linear equation in two variables, typically denoted as x and y, has a general form as follows:

$$Ax + By = C$$

where; A, B, and C are integers; A is typically a positive integer (though it can be negative); and A, B, and C share no common factors greater than 1.

The standard form allows for easy comparison of two linear equations and is conducive to solving systems of linear equations through methods like substitution or elimination.

Ø **Example:** Consider the following two linear equations:

$$2x - 3y = 7$$

$$4x + 6y = 12$$

Both equations are standard, making comparing coefficients and determining their relationship straightforward.

4.2.3. The Slope-Intercept Form

The slope-intercept form of a linear equation is a versatile and widely used representation of lines in the Cartesian coordinate system.

It offers an intuitive way to understand and describe lines, making it an essential tool in various mathematical and real-world contexts (Crocetto & Russo, 1994).

4.2.3.1. The Equation

The slope-intercept form of a linear equation is written as:

$$y = mx + b$$

In this equation, y and x are variables representing the coordinates of points on the line; m is the slope of the line; and b is the y-intercept.

4.2.3.2. Components

1. **Slope** (m): The slope of a line indicates how steeply it inclines or declines. It quantifies the rate at which the y-coordinate changes concerning the x-coordinate. Mathematically, the slope is defined as the change in y $(\text{Ä}y)$ divided by the change in x $(\text{Ä}x)$, often denoted as:

$$m = \frac{Äy}{Äx}$$

The slope determines both the direction and steepness of the line. A positive slope means the line slants upward from left to right, while a negative slope indicates a downward slant. A horizontal line has a zero slope, and a vertical line has an undefined slope (Francoeur et al., 2009).

Ø **Example 1:** Consider the line $(y=3x+2)$. Here, the slope is $(m=3)$. It implies that for every one-unit increase in the x-coordinate, the y-coordinate increases by two units (Figure 4.5).

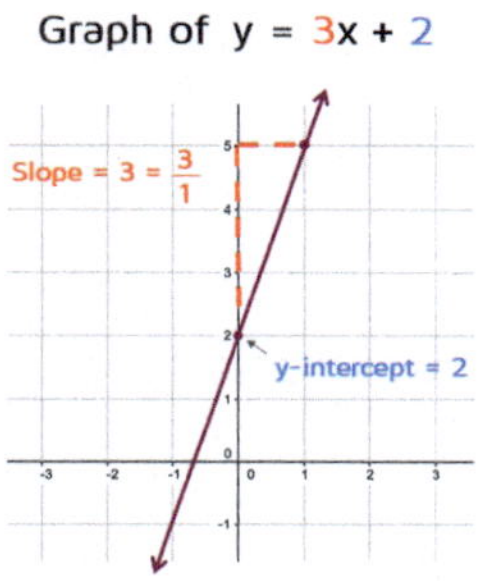

Figure 4.5. Schematic of the slop intercept form.

Source: All Math, Creative Commons License.

2. **Y-Intercept** (b): The y-intercept is the point at which the line intersects the y-axis, where the x-coordinate is zero. It is represented as the coordinate $(0,b)$ on the line.

Ø **Example 2:** In the equation $(y=2x+3)$, the y-intercept is $(b=3)$. It means that the line crosses the y-axis at the point $(0,3)$.

4.2.3.3. Graphical Interpretation

The graphical interpretation of the slope-intercept form is particularly intuitive. The slope (m) tells us how to move from one point on the line to another. Starting from the y-intercept $(0,b)$, we move horizontally to the right by one unit $(Äx=1)$ and then vertically up or down by (m) units $(Äy=m)$. It allows us to generate additional points on the line (Kaburaki et al., 2018).

- Ø **Example 3:** For the equation $(y=2x+3)$, we start at the y-intercept $(0,3)$. To find another point, we move one unit to the right $(\text{Ä}x=1)$ and then two units up $(\text{Ä}y=2)$ to reach the point $(1,5)$. We can continue this process to plot more points.

4.2.4. Point-Slope Form

The point-slope form is an alternative and efficient way to represent linear equations, particularly when you know a point on the line and its slope. This form offers mathematical precision and is highly practical for solving problems involving lines on a plane.

4.2.4.1. The Equation

The point-slope form of a linear equation is given by:

$$y-y_1=m(x-x_1)$$

In this equation, y and x are variables representing the coordinates of points on the line; m is the slope of the line; and (x_1,y_1) are the coordinates of a known point on the line.

4.2.4.2. Components

1. **Slope (m):** Just like in the slope-intercept form, the slope (m) represents the rate of change of the line. It quantifies how the y-coordinate changes concerning the x-coordinate.

2. **Point (x_1,y_1):** Unlike the slope-intercept form, the point-slope form directly includes the coordinates of a known point on the line. This point is used to anchor the line in the plane (Cole et al., 1993).
3. **The Point-Slope Formula:** The point-slope form can be derived from the concept of slope. The slope (m) between two points (x_1,y_1) and (x_2,y_2) is defined as:

$$m = \frac{y_2 - y_1}{x_2 - x_1}$$

Using this definition, we can rewrite the slope as:

$$m = \frac{y - y_1}{x - x_1}$$

Now, if we multiply both sides by $(x - x_1)$, we obtain the point-slope form:

$$y - y_1 = m(x - x_1)$$

This equation allows us to precisely represent a line, given a point on the line and its slope.

4.2.4.3. Graphical Interpretation

Graphically, the point-slope form provides a straightforward way to draw a line. Start with the known point. (x_1, y_1) and use the slope (m) to determine how the line slants and extends from that point (Urankar, 1990).

- Ø **Example 1:** Suppose we have an equation $y - 1 = 2(x - 2)$. Initially, we identify the slope as (2) and observe that the line passes through the point $(2,1)$. Using this information, we can systematically generate additional points on the line by applying the slope-intercept form. For each one-unit movement to the right $(x - 2 = 0)$, we correspondingly move two units up $2(x-2) = 2(0) = 0$, which leads us to the point $(3,3)$. Continuing this pattern, with another one-unit shift to the right $(x - 2 = 1)$, we move two units up $2(x-2) = 2(1) = 2$, resulting in the point $(4,5)$. It allows us to systematically plot more points on the line, consistently moving one unit to the right and two units up from the preceding point (Birgin, 2012) (Figure 4.6).

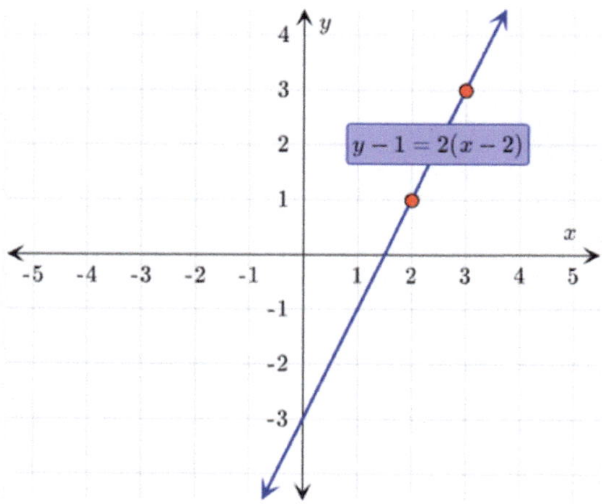

Figure 4.6. Illustration of the example of point-slop form.

Source: Mathleaks, Creative Commons License.

4.2.5. Parallel and Perpendicular Lines

In the Cartesian plane, understanding the relationships between lines is essential. Two crucial relationships are parallel lines and perpendicular lines. Specific characteristics and equations define these relationships and play a fundamental role in geometry and various applications.

Tip

When dealing with parallel lines, remember that they have the same slope. For perpendicular lines, their slopes are negative reciprocals of each other. This relationship simplifies finding equations for these lines.

4.2.5.1. Parallel Lines

Parallel lines are two or more lines that run in the same direction and never intersect, regardless of how far they are extended in both directions. They maintain a constant separation or distance between them (Sablin, 2020).

The following are the characteristics of parallel lines:

1. **Equal Slopes:** One defining characteristic of parallel lines is that they have the same slope ( ) If two lines are parallel, they share the same rate of y change concerning x.

$$m_1 = m_2$$

where; m_1 and m_2 represent the slopes of the two parallel lines.

2. **Different Y-Intercepts:** Parallel lines can have different Y-intercepts (b). The y-intercept is where a line crosses the y-axis (Long et al., 2013).

$$b_1 \neq b_2$$

The subscripts (1) and (2) denote the two lines' y-intercepts.

Following are the equations of parallel lines:

Given a line with slope (m) and y-intercept (b), the equation of this line is:

$$y = mx + b$$

For parallel lines, the equation remains the same, and only the y-intercept changes:

$$y = mx + b_1$$

$$y = mx + b_2$$

where; b_1 and b_2 are the different y-intercepts of the parallel lines.

Ø **Example 1:** Consider the lines $(y = 2x + 3)$ and $(y = 2x - 1)$. These lines are parallel because they have the same slope $(m = 2)$ but different y-intercepts $(b_1 = 3)$ and $(b_2 = -1)$.

4.2.5.2. Perpendicular Lines

Perpendicular lines are two lines that intersect at a right angle, forming four right angles at their point of intersection.

The following are the characteristics of perpendicular lines:

1. **Negative Reciprocal Slopes:** The key characteristic of perpendicular lines is that their slopes are negative reciprocals. If the slope of one line is (m), the slope of a line perpendicular to it is $\left(-\frac{1}{m}\right)$.

$$m_1 \cdot m_2 = -1$$

where; m_1 and m_2 represent the slopes of the two perpendicular lines (Ryckaert et al., 1977).

Following is the equation of perpendicular lines:

Given a line with slope (m) and y-intercept (b), the equation of this line is:

$$y = mx + b$$

For perpendicular lines, the equation changes to:

$$y = -\frac{1}{m}x + b_1$$

where; (b_1) is the y-intercept of the perpendicular line.

Ø **Example 2:** Let's consider the lines $(y = 2x + 3)$ and $\left(y = -\frac{1}{2}x + 1\right)$. These lines are perpendicular because the slope of the first line is $(\quad 2)$, and the slope of the second line is the negative reciprocal of (m_1), which is $\left(m_2 = -\frac{1}{2}\right)$.

4.3. EQUATIONS OF CURVES

In coordinate geometry, conic sections are some of the most captivating and mathematically rich curves. These curves include circles, ellipses, hyperbolas, and parabolas. Each possesses unique properties, equations, and geometric characteristics, making them a profound topic of study. This explores the equations of these remarkable curves, revealing their essential features, standard forms, eccentricity, and the intriguing concepts of focus and directrix (Monari Martinez & Neodo, 2022).

4.3.1. Conic Sections

Conic sections are the result of slicing a double-napped cone with a plane. We obtain different conic sections depending on the angle and position of the intersecting plane. The primary conic sections are circles, ellipses, hyperbolas, and parabolas, each with its distinctive equation (Figure 4.7).

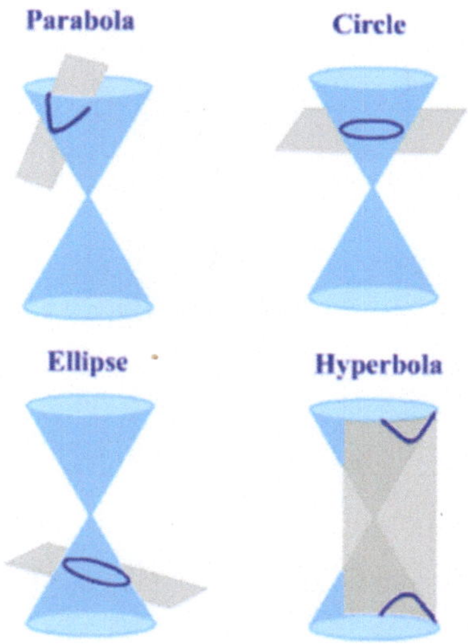

Figure 4.7. Illustration of the various conic sections.

Source: Varsity Tutors, Creative Commons License.

4.3.1.1. Circles

A circle is a set of points equidistant from a fixed point called the center. The distance from the center to any point on the circle is called the radius. In the Cartesian plane with the center (h,k) and radius (r), the standard equation of a circle is:

$$(x-h)^2+(y-k)^2=r^2$$

This equation defines a circle centered at (h,k) with radius (r). It is a fundamental shape in geometry, with applications ranging from geometry problems to engineering and physics (Volodymyr & Anatolii, 2018).

4.3.1.2. Ellipses

An ellipse is another fascinating conic section. It consists of all points where the sum of the distances to two fixed points, known as foci, remains constant. The distance between the foci is $(2c)$, and the length of the major axis is $(2a)$. The standard equation for an ellipse with center (h,k) and major and minor axes lengths $(2a)$ and $(2b)$, respectively, are:

$$\frac{(x-h)^2}{a^2}+\frac{(y-k)^2}{b^2}=1$$

The eccentricity (e) of an ellipse measures its "roundness." It is defined as $\left(e=\frac{c}{a}\right)$, where (c) is the distance between the center and a focus (Chopra et al., 2022).

4.3.1.3. Hyperbolas

Hyperbolas have two branches, consisting of all points where the difference of the distances to two fixed points, the foci, is constant. The distance between the foci is $(2c)$, and the distance between the vertices is $(2a)$. The standard equations for the transverse axis along the x-axis and y-axis, respectively, are:

$$\frac{(x-h)^2}{a^2}-\frac{(y-k)^2}{b^2}=1$$

- **For a Horizontal Transverse Axis:**

$$\frac{(y-k)^2}{b^2}-\frac{(x-h)^2}{a^2}=1$$

- **For a Vertical Transverse Axis:**

The eccentricity (e) of a hyperbola, as in ellipses, is given by $\left(e=\frac{c}{a}\right)$.

4.3.1.4. Parabolas

A parabola is a curve consisting of all points equidistant to a fixed point (the focus) and a fixed line (the directrix). The distance between the vertex (the point on the parabola closest to the focus) and the focus is (p), and the standard equations for a parabola with the vertex at (h,k) are:

$(x-h)^2=4p(y-k)$ (vertical axis)

$(y-k)^2=4p(x-h)$ (horizontal axis)

Here, (p) is half the distance between the focus and the directrix. The eccentricity of a parabola is $(e=1)$, which signifies its unique nature among conic sections (Han et al., 2018).

4.3.2. General Equation of a Circle

While the standard equation of a circle describes circles centered at the origin, a more general form is needed to describe circles with centers at arbitrary points. The general equation of a circle is:

$$(x-h)^2+(y-k)^2=r^2$$

where; (h,k) represents the circle's center; and (r) is the radius. This equation enables the description of circles located anywhere in the Cartesian plane.

4.3.3. Standard Forms of Conic Sections

Conic sections have specific standard forms for ease of recognition and analysis. These forms help identify each curve's shape, orientation, and dimensions.

1. **Circles:** The standard form for a circle is $(x-h)^2+(y-k)^2=r^2$, which is centered at (h,k) with radius (r).
2. **Ellipses:** The standard form for an ellipse with a horizontal major axis is $\frac{(x-h)^2}{a^2}+\frac{(y-k)^2}{b^2}=1$, and with a vertical major axis, it's $\frac{(y-k)^2}{b^2}+\frac{(x-h)^2}{a^2}=1$. Here, (h,k) is the center; $(2a)$ is the major axis length; and $(2b)$ is the minor axis length (Ozcelik & Acarturk, 2011).
3. **Hyperbolas:** For a hyperbola with a horizontal transverse axis, the standard form is $\left(\frac{(x-h)^2}{a^2}-\frac{(y-k)^2}{b^2}=1\right)$. For a vertical transverse

axis, it's $\left(\frac{(y-k)^2}{b^2}-\frac{(x-h)^2}{a^2}=1\right)$. The center is (h,k),$(2a)$ is the distance between vertices, and $(2b)$ is the distance between foci.

4. **Parabolas:** These have two standard forms: $(x-h)^2=4p(y-k)$ (vertical axis) and $(y-k)^2=4p(x-h)$ (horizontal axis). The vertex is (h,k), and (p) represents half the distance between the focus and directrix.

4.3.4. Eccentricity

Eccentricity (e) is a measure of how "stretched" or "flattened" a conic section is. For ellipses and hyperbolas, it is defined as $\left(e=\frac{c}{a}\right)$, where (c) is the distance between the center and a focus, and (a) is half the length of the major axis. For circles $(e=0)$ and parabolas $(e=1)$, the eccentricity takes on specific values, while for ellipses and hyperbolas, (e) is always greater than or equal to (1).

Remember

Eccentricity is a crucial parameter for conic sections. It defines the shape of the conic. Remember that for circles, the eccentricity is always 0, for ellipses it's between 0 and 1, for parabolas it's 1, and for hyperbolas, it's greater than 1.

4.3.5. Focus and Directrix

The concepts of focus and directrix are pivotal in defining conic sections:

1. **Ellipse Focus and Directrix:** For an ellipse, there are two foci, each located at a distance of (c) from the center. The major axis bisects the foci. The directrices are two horizontal lines equidistant from the center, lying on either side (Rote, 1991).
2. **Hyperbola Focus and Directrix:** In a hyperbola, there are also

two foci, each located at a distance of (c) from the center. The transverse axis bisects the foci. The directrices are two horizontal lines equidistant from the center.

3. **Parabola Focus and Directrix:** A parabola has one focus, equidistant from the vertex and the directrix. The focus is located at a distance (p) along the axis of symmetry from the vertex. The directrix is a horizontal line at a distance (p) from the vertex.

4.4. COORDINATE GEOMETRY IN THREE DIMENSIONS

Coordinate geometry, a powerful branch of mathematics, extends its reach into three dimensions, allowing us to explore space with precision and clarity. In this section, we will learn about plotting points, calculating distances and midpoints, and deriving equations of lines and planes in three-dimensional space. This expansion into the third dimension opens new vistas of mathematical exploration and practical applications (Crouse & Skufca, 2018).

4.4.1. 3D Coordinate Systems

In 3D coordinate geometry, we build upon the familiar Cartesian plane by adding dimension. Instead of the usual two axes (x and y), we introduce a third axis, often labeled z. It creates a 3D coordinate system, sometimes referred to as XYZ coordinates.

In a 3D coordinate system, points are represented by ordered triples (x, y, z), where; 'x' represents the horizontal position on the x-axis; 'y' represents the horizontal position on the y-axis; and 'z' represents the vertical position on the z-axis.

To plot a point in 3D space, locate its coordinates along the respective axes. For example, the point (2, 3, 4) would be located two units to the right (x-axis), three units up (y-axis), and four units in front (z-axis) from the origin (0, 0, 0) (Wesolowsky & Love, 1971).

4.4.2. Distance and Midpoint in 3D

Extending the concepts from 2D to 3D, we can derive formulas for distance and midpoint in three dimensions.

4.4.2.1. Distance Formula in 3D

To calculate the Euclidean distance between two points (x_1, y_1, z_1) and (x_2, y_2, z_2) in 3D space, you can use the 3D distance formula:

$$d = \sqrt{(x_2 - x_1)^2 + (y_2 - y_1)^2 + (z_2 - z_1)^2}$$

Each term represents the difference between the corresponding coordinates, squared, and summed under a square root.

4.4.2.2. Midpoint Formula in 3D

Finding the midpoint between two points in 3D space (x_1, y_1, z_1) and (x_2, y_2, z_2) is equally straight-forward. The 3D midpoint formula is:

$$M = \left(\frac{x_1 + x_2}{2}, \frac{y_1 + y_2}{2}, \frac{z_1 + z_2}{2} \right)$$

It averages the coordinates of the two points x, y, and z to find the midpoint.

4.4.3. Equations of Lines and Planes in 3D

Just as we can define lines and planes in 2D coordinate geometry, we can extend these concepts to 3D space.

4.4.3.1. Equation of a Line in 3D

A line in 3D space can be defined as the set of all points (x, y, z) that satisfy the equations:

$$x = x_1 + at$$

$$y = y_1 + bt$$

$$z = z_1 + ct$$

here; x_1, y_1, z_1 is a known point on the line; $a,b,$ and c are the direction ratios that determine the line's direction; t is a parameter that varies over real numbers and generates all points on the line (Ishibuchi et al., 2015).

4.4.3.2. Equation of a Plane in 3D

An equation of the form can represent a plane in 3D space:

$$Ax+By+Cz=D$$

here; A, B, C are the coefficients of the plane's normal vector, which determines the plane's orientation; and (D) is a constant.

Ø **Example: Applying 3D Coordinate Geometry**

- **Problem 1: Flight Path:** An airplane takes off from an airport at coordinates (3, 2, 1) and lands at another airport with coordinates (10, 12, 15). Calculate the Euclidean distance traveled by the airplane during its flight.
- **Solution:**

Apply the 3D distance formula:

$$d=\sqrt{(10-3)^2+(12-2)^2+(15-1)^2}$$

$$d=\sqrt{7^2+10^2+14^2}$$

$$d=\sqrt{49+100+196}$$

$$d=\sqrt{345}$$

The airplane traveled $\left(\sqrt{345}\right)$ units during its flight.

- **Problem 2: Architectural Design:** An architect is designing a building with the following features: one corner is at coordinates (2, 4, 6), another corner is at (8, 6, 10), and a third corner is at (5, 8, 7). Find the equation of the plane formed by these three points (Oszczak & Sitnik, 2013).
- **Solution:**

First, calculate two direction vectors:

- **Vector 1:** $(8-2,6-4,10-6)=(6,2,4)$
- **Vector 2:** $(5-2,8-4,7-6)=(3,4,1)$

Now, find the normal vector by taking the cross-product of these vectors:

- **Normal Vector:** $\left((6i+2j+4k)\times(3i+4j+k)=2i-14j+14k\right)$

Now, use the point-normal form of the equation of a plane:

$$2(x-2)-14(y-4)+14(z-6)=0$$

Simplify:

$$2x-14y+14z-28+56-84=0$$

$$2x-14y+14z-56=0$$

It is the equation of the plane formed by the given points.

- **Problem 3: Robotic Arm:** A robotic arm starts at coordinates (1, 2, 3) and extends in the direction of vector $(2,3,1)$. Find the coordinates of a point on the robotic arm 5 units away from the starting point (Tsai, 1987).
- **Solution:** Let (x, y, z) be the coordinates of the desired point on the robotic arm. We know that the direction ratios of the robotic arm are $(a=2)$,$(b=3)$, and $(c=1)$, and we want the distance to be 5 units. Using the 3D equation of a line:

$$x=1+2t$$

$$y=2+3t$$

$$z=3+t$$

We can use the distance formula to find (t):

$$5=\sqrt{(x-1)^2+(y-2)^2+(z-3)^2}$$

$$5=\sqrt{\left[(2t)^2+(3t)^2+(t)^2\right]}$$

$$5=\sqrt{14t^2}$$

Solve for (t):

$$25=14t^2$$

$$t^2=\frac{25}{14}$$

$$t=\pm\sqrt{\frac{25}{14}}$$

Now, substitute (t) back into the equations:

For $\left(t=\sqrt{\frac{25}{14}}\right)$:

$$x=1+2\sqrt{\frac{25}{14}}$$

$$y=2+3\sqrt{\frac{25}{14}}$$

$$z=3+\sqrt{\frac{25}{14}}$$

For

$$\left(t=-\sqrt{\frac{25}{14}}\right):$$

$$x=1-2\sqrt{\frac{25}{14}}$$

$$y=2-3\sqrt{\frac{25}{14}}$$

$$z=3-\sqrt{\frac{25}{14}}$$

These are the coordinates of the two points on the robotic arm 5 units away from the starting point.

In this section, we have discussed 3D coordinate geometry, where we extended the concepts of plotting points, calculating distances and midpoints, and deriving equations of lines and planes into three dimensions. These mathematical tools are invaluable in various fields, from engineering and physics to computer graphics and architecture.

Through example problems, we have seen how these concepts can be applied to solve real-world challenges, demonstrating the practical significance of 3D coordinate geometry (Nemirovsky et al., 2012).

4.5. EXAMPLES OF DISTANCE AND MIDPOINT

Coordinate geometry or Cartesian geometry bridges algebra and geometry, enabling us to represent geometric shapes and solve problems using algebraic equations. Two fundamental concepts in this field are distance and midpoint. We will delve into these concepts, exploring the Euclidean distance formula, the midpoint formula, and their real-world applications, and solving example problems to illuminate their significance.

Let's work through some example problems to illustrate the practical use of the distance and midpoint formulas.

Ø **Problem 1: Navigating a City**

You are in a city and want to reach a famous landmark at coordinates (3, 5). You are currently at (1, 2). How far do you need to travel to reach the landmark?

- **Solution:** Use the Euclidean distance formula:

$$d = \sqrt{(3-1)^2 + (5-2)^2} = \sqrt{2^2 + 3^2} = \sqrt{4+9} = \sqrt{13}$$

So, you need to travel $\left(\sqrt{13}\right)$ units to reach the landmark.

Ø **Problem 2: Finding the Midpoint:** You have a rectangular garden with one corner at (2, 4) and the opposite corner at (8, 10). Where is the midpoint of the garden, which represents the ideal spot for a decorative fountain?

- **Solution:** Apply the midpoint formula:

$$M = \left(\frac{2+8}{2}, \frac{4+10}{2}\right) = (5,7)$$

The midpoint of the garden is (5, 7), which is the ideal spot for the fountain.

Ø **Problem 3: Flight Planning:** You are planning a flight from an airport at (1, 2) to another airport at (9, 8). How far is the destination airport from your current location?

Solution: Use the Euclidean distance formula:

$$d = \sqrt{(9-1)^2 + (8-2)^2} = \sqrt{8^2 + 6^2} = \sqrt{64+36} = \sqrt{100} = 10$$

The destination airport is 10 units away from your current location.

Ø **Problem 4: Architecture and Construction:** An architect is designing a rectangular room with one corner at (3, 2) and the opposite corner at (9, 6). What are the coordinates of the room's midpoint, which will be the focal point of the room's design?

- **Solution:** Apply the midpoint formula:

$$M = \left(\frac{3+9}{2}, \frac{2+6}{2}\right) = (6,4)$$

The coordinates of the room's midpoint are (6, 4), which will serve as the focal point of the room's design.

Ø **Problem 5: Geocaching Adventure:** You are on a geocaching adventure, and your GPS receiver indicates that the cache is at coordinates (7, 3). Your current location is (4, 1). How far are you from the cache?

- **Solution:** Use the Euclidean distance formula:

$$d = \sqrt{(7-4)^2 + (3-1)^2} = \sqrt{3^2 + 2^2} = \sqrt{9+4} = \sqrt{13}$$

You are $\left(\sqrt{13}\right)$ units away from the cache's location.

These example problems demonstrate the practical significance of distance and midpoint calculations in various real-world scenarios, from navigation to design and beyond (Kirmaci, 2004).

In conclusion, coordinate geometry, with its distance and midpoint formulas, is vital in solving real-world problems. Whether navigating a city, designing a room, planning a flight, or embarking on a geocaching adventure, these formulas provide the mathematical tools to bridge the gap between theory and practice, ensuring precision and accuracy in your endeavors.

4.6. APPLICATIONS OF COORDINATE GEOMETRY

Coordinate geometry, a powerful branch of mathematics, bridges abstract mathematical concepts and practical real-world applications. This section explores the multifaceted applications of coordinate geometry in various fields, from physics and engineering to computer science and calculus.

4.6.1. Physics

Coordinate geometry plays a fundamental role in physics, particularly in describing the motion of objects in space. For instance, when analyzing the trajectory of a projectile, such as a thrown ball or a rocket, physicists use 2D or 3D coordinates to track its position over time. The equations of motion involve coordinates, velocities, and accelerations, enabling accurate predictions of an object's path. Consider a ball launched from ground level at coordinates (0, 0) with an initial velocity of (20 m/s) in the direction of 45° above the horizontal axis (x-axis). Using coordinate geometry, you can derive equations that describe its height above the ground and horizontal distance traveled at any time (t). These equations are essential for understanding and predicting the ball's behavior.

4.6.2. Engineering

Engineers rely heavily on coordinate geometry for designing and analyzing structures and systems. Civil engineers, for example, use coordinates to plan the layout of roads, bridges, and buildings. They employ coordinate systems to ensure precise measurements and alignments during construction projects. In structural engineering, coordinate geometry helps engineers' model and analyze the load-bearing capacity of materials and structures. By defining the coordinates of key points and applying principles of physics, they can simulate how structures will respond to various forces and stresses (Kirmaci & Özdemir, 2004).

4.6.3. Computer Science

Computer graphics, a vital field in computer science, relies on coordinate geometry to create, manipulate, and render images and animations in 2D and 3D space. Graphics software, video games, and simulations use coordinates to represent objects, cameras, and lighting. Consider a 3D video game where a virtual camera moves through a digital world. The position and orientation of the camera are described using coordinates and vectors, allowing the game engine to render scenes from the camera's perspective in real time. Moreover, collision detection algorithms use coordinates to determine whether objects in the virtual world intersect (Dragomir, 2000).

4.6.4. Analytical Geometry in Calculus

Coordinate geometry is tightly integrated with calculus, providing a geometric interpretation of mathematical concepts and facilitating problem-solving.

4.6.4.1. Tangents and Derivatives

In calculus, the derivative of a function at a point represents the tangent line's slope to the function's graph at that point. Coordinate geometry enables us to find the equation of this tangent line. For example, to find the tangent line to the curve $(y = x^2)$ at the point (2, 4), we use the slope-intercept form $(y - y_1) = m(x - x_1)$, where (m) is the derivative of the function at (x[illegible]) (Gaitonde & Visbal, 1998).

4.6.4.2. Integrals and Area

The concept of integration, central to calculus, involves finding the area under curves. Coordinate geometry allows us to visualize this area. For instance, to find the area under the curve $(y = x^2)$ between $(x=1)$ and $(x=3)$, we can geometrically interpret this as the area between the curve, the x-axis, and the vertical lines $(x=1)$ and $(x=3)$. Using coordinate geometry, we can compute this area precisely.

4.6.5. Vector Geometry

Coordinate geometry provides the foundation for vector geometry, a branch of mathematics crucial in physics, engineering, and computer science.

Did You Know?

Coordinate geometry plays a significant role in vector geometry. Vectors are used in various fields, including physics, engineering, and computer graphics, to represent quantities with both magnitude and direction. They're a powerful tool for solving real-world problems involving forces, motion, and more.

4.6.5.1. Vectors in 2D and 3D

In 2D and 3D space, vectors are represented using coordinates. A vector has both magnitude and direction; coordinate differences determine its components. For example, in 2D, a vector $(\vec{v})$ can be expressed as $\vec{v} = (a, b)$, where (a) and (b) represent the differences in the x and y coordinates, respectively (Tamura & Tanno, 2008).

4.6.5.2. Dot and Cross Products

The dot product and cross product of vectors are essential operations in vector geometry. They are used to calculate angles between vectors, determine orthogonality, and find vector projections.

In 3D space, the dot product of two vectors $\vec{u} = (a, b, c)$ and $\vec{v} = (x, y, z)$ is given by:

$$\vec{u} \cdot \vec{v} = ax + by + cz$$

The cross-product of two vectors $(\vec{u})$ and $(\vec{v})$ yields a vector orthogonal to both $(\vec{u}) and (\vec{v})$ and is defined as:

$$\{\vec{u}\} \times \{\vec{v}\} = (bc - cy, cz - ac, ax - bx)$$

4.6.6. Computer Graphics and Modeling

Coordinate geometry is the backbone of computer graphics and 3D modeling, bringing virtual worlds to life.

4.6.6.1. 3D Modeling

In 3D modeling software, objects are constructed and manipulated using coordinates and geometric shapes. Each vertex of a 3D model is defined by its position in 3D space. This information is stored as coordinates (x, y, z), allowing the software to render complex 3D scenes (Adams & Pneuman, 1976).

4.6.6.2. Transformation Matrices

Coordinate transformations, such as translation, rotation, and scaling, are fundamental in computer graphics. These transformations are represented using matrices, making it possible to move, rotate, and resize objects seamlessly in a 3D environment. Matrices enable efficient computations of these transformations, which are then applied to the coordinates of vertices.

4.6.6.3. Rendering and Ray Tracing

Rendering engines use coordinate geometry to simulate the interaction of light with surfaces in a 3D scene. Through ray tracing techniques, they calculate the paths of light rays as they interact with objects, calculating shadows, reflections, and refractions accurately.

In summary, coordinate geometry is a versatile and indispensable tool across various fields, allowing mathematicians, scientists, engineers, and computer scientists to solve complex problems and model real-world phenomena with precision. Its applications extend from the physical sciences, where it aids in understanding the behavior of objects, to the realms of calculus and vector geometry, enriching our mathematical toolkit. In computer science and graphics, it forms the basis of creating immersive virtual environments and simulations. Thus, coordinate geometry is a foundation of modern mathematics with far-reaching implications for understanding the world and our ability to harness mathematical concepts for practical use (Setzmann & Wagner, 1991).

Activity 4.1

Imagine you're an architect tasked with designing a new skyscraper. Using the principles of coordinate geometry, explain how you would determine the placement of windows on each floor to maximize natural light while considering the structural integrity of the building.

4.7. SUMMARY

This chapter introduces coordinate geometry, a practical tool for understanding shapes and lines using numbers. It starts with the basics, like the Cartesian plane and plotting points and lines. You'll learn about line equations and their different forms, making it easier to understand their properties. The chapter also explores curves, like circles and conic sections, showing how to work with their equations. In 3D, it introduces 3D coordinate systems and how to find distances and equations for lines and planes. Practical applications span physics, engineering, computer science, and more, making coordinate geometry an essential skill in various fields.

REFERENCES

1. Adams, J., & Pneuman, G. W., (1976). A new technique for determining coronal magnetic fields: A fixed mesh solution to Laplace's equation using line-of-sight boundary conditions. *Solar Physics*, *46*, 185–203.
2. Agarwal, P., Vivas-Cortez, M., Rangel-Oliveros, Y., & Ali, M. A., (2022). New Ostrowski type inequalities for generalized s-convex functions with applications to some special means of real numbers and to midpoint formula. *AIMS Mathematics*, *7*(1), 1429–1444.
3. Alam, C. N., Manaf, K., Atmadja, A. R., & Aurum, D. K., (2016). Implementation of haversine formula for counting event visitor in the radius based on Android application. In: *2016 4th International Conference on Cyber and IT Service Management* (pp. 1–6). IEEE.
4. Bakuła, M., (2013). Study of reliable rapid and ultrarapid static GNSS surveying for determination of the coordinates of control points in obstructed conditions. *Journal of Surveying Engineering*, *139*(4), 188–193.
5. Birgin, O., (2012). Investigation of eighth-grade students' understanding of the slope of the linear function. *Bolema: Boletim de Educação Matemática*, *26*, 139–162.
6. Bliss, R. L., Katz, J. N., Wright, E. A., & Losina, E., (2012). Estimating proximity to care: are straight line and zipcode centroid distances acceptable proxy measures? *Medical Care*, *50*(1), 99.
7. Buhr, E., Günther-Kohfahl, S., & Neitzel, U., (2003). Accuracy of a simple method for deriving the presampled modulation transfer function of a digital radiographic system from an edge image. *Medical Physics*, *30*(9), 2323–2331.
8. Carrasco-Navarro, R., Luque-Vega, L. F., Nava-Pintor, J. A., Guerrero-Osuna, H. A., Carlos-Mancilla, M. A., & Castañeda-Miranda, C. L., (2022). MEIoT 2D-CACSET: IoT two-dimensional cartesian coordinate system educational toolkit align with educational mechatronics framework. *Sensors*, *22*(13), 4802.
9. Chaplot, V., Darboux, F., Bourennane, H., Leguédois, S., Silvera, N., & Phachomphon, K., (2006). Accuracy of interpolation techniques for the derivation of digital elevation models in relation to landform types and data density. *Geomorphology*, *77*(1–2), 126–141.
10. Chopra, G., & Pal, A. K., (2010). An improved image compression algorithm using binary space partition scheme and geometric wavelets. *IEEE Transactions on Image Processing*, *20*(1), 270–275.

11. Chopra, S., Luthra, N., Kathuria, S., Saggar, K., Gupta, S., & Singh, T., (2022). Optimal positioning of right internal jugular venous catheter: A randomized study comparing modified Peres' height formula and distance between insertion point and right third intercostal space. *Indian Journal of Anesthesia*, *66*(8), 585–590.
12. Cole, G. K., Nigg, B. M., Ronsky, J. L., & Yeadon, M. R., (1993). *Application of the Joint Coordinate System to Three-Dimensional Joint Attitude and Movement Representation: A Standardization Proposal*, 115(4A), 344-349.
13. Crocetto, N., & Russo, P., (1994). Helmert's projection of a ground point onto the rotational reference ellipsoid in topocentric cartesian coordinates. *Bulletin Géodésique*, *69*, 43–48.
14. Crofton, M. W., (1868). VII: On the theory of local probability, applied to straight lines drawn at random in a plane; the methods used being also extended to the proof of certain new theorems in the integral calculus. *Philosophical Transactions of the Royal Society of London*, *158*, 181–199.
15. Crouse, D., & Skufca, J., (2018). On the nature of discrete space-time: Part 1: The distance formula, relativistic time dilation and length contraction in discrete space-time. *arXiv Preprint arXiv:1803.03126*.
16. De Sadeleer, C., Piepsz, A., & Ham, H. R., (2007). Checking the consistency of the two blood samples slope–intercept method for estimating GFR using the single blood sample formula in children. *Nuclear Medicine Communications*, *28*(1), 49–54.
17. Dragomir, S. S., (2000). On the midpoint quadrature formula for mappings with bounded variation and applications. *Kragujevac J. Math*, *22*(1), 13–19.
18. Earnest, D., (2015). From number lines to graphs in the coordinate plane: Investigating problem solving across mathematical representations. *Cognition and Instruction*, *33*(1), 46–87.
19. Ebenezer, D. D., & Stepanishen, P. R., (1991). Wave-vector-time domain and Kirchhoff integral equation methods to determine the transient acoustic radiation loading on circular cylinders. *The Journal of the Acoustical Society of America*, *89*(6), 2532–2544.
20. Fan, Z., Wang, T., Cheng, Z., Li, G., & Gu, F., (2017). An improved multiobjective particle swarm optimization algorithm using minimum distance of point to line. *Shock and Vibration*, *2017*.

21. Faria, T., & Magalhães, L. T., (1995). Normal forms for retarded functional differential equations and applications to Bogdanov-Takens singularity. *Journal of Differential Equations*, *122*(2), 201–224.
22. Finch, N. C., Syme, H. M., Elliott, J., Peters, A. M., Gerritsen, R., Croubels, S., & Heiene, R., (2011). Glomerular filtration rate estimation by use of a correction formula for slope-intercept plasma iohexol clearance in cats. *American Journal of Veterinary Research*, *72*(12), 1652–1659.
23. Fleming, J. S., (2007). An improved equation for correcting slope–intercept measurements of glomerular filtration rate for the single exponential approximation. *Nuclear Medicine Communications*, *28*(4), 315–320.
24. Fleming, J. S., Persaud, L., & Zivanovic, M. A., (2005). Quality control of slope–intercept measurements of glomerular filtration rate using single-sample estimates. *Nuclear Medicine Communications*, *26*(8), 737–742.
25. Francoeur, M., Mengüç, M. P., & Vaillon, R., (2009). Solution of near-field thermal radiation in one-dimensional layered media using dyadic Green's functions and the scattering matrix method. *Journal of Quantitative Spectroscopy and Radiative Transfer*, *110*(18), 2002–2018.
26. Gaitonde, D. V., & Visbal, M., (1998). *High-Order Schemes for Navier-Stokes Equations: Algorithm and Implementation into FDL3DI*. Air Vehicles Directorate, Air Force Research Laboratory, Air Force Materiel Command,1(1), 1-16.
27. Gavazzi, G., Boselli, A., Scodeggio, M., Pierini, D., & Belsole, E., (1999). The 3D structure of the Virgo cluster from H-band fundamental plane and Tully—Fisher distance determinations. *Monthly Notices of the Royal Astronomical Society*, *304*(3), 595–610.
28. Güngör, N., Özyürek, M., Güçlü, K., Çekiç, S. D., & Apak, R., (2011). Comparative evaluation of antioxidant capacities of thiol-based antioxidants measured by different *in vitro* methods. *Talanta*, *83*(5), 1650–1658.
29. Hamson, R. M., (1985). The theoretical responses of vertical and horizontal line arrays to wind-induced noise in shallow water. *The Journal of the Acoustical Society of America*, *78*(5), 1702–1712.
30. Han, D., Nie, H., Chen, J., & Chen, M., (2018). Dynamic obstacle avoidance for manipulators using distance calculation and discrete

detection. *Robotics and Computer-Integrated Manufacturing*, *49*, 98–104.

31. Ho, J. C., Russel, K. C., & Davis, J., (2014). The challenge of global water access monitoring: Evaluating straight-line distance versus self-reported travel time among rural households in Mozambique. *Journal of Water and Health*, *12*(1), 173–183.
32. Ikemoto, T., & Takai, K., (2000). A new linearized formula for the law of total effective temperature and the evaluation of line-fitting methods with both variables subject to error. *Environmental Entomology*, *29*(4), 671–682.
33. Ishibuchi, H., Masuda, H., Tanigaki, Y., & Nojima, Y., (2015). Modified distance calculation in generational distance and inverted generational distance. In: *Evolutionary Multi-Criterion Optimization: 8th International Conference, EMO 2015, Guimarães, Portugal, Proceedings, Part II 8* (pp. 110–125). Springer International Publishing.
34. Jacques, S. L., (1998). Light distributions from point, line, and plane sources for photochemical reactions and fluorescence in turbid biological tissues. *Photochemistry and Photobiology*, *67*(1), 23–32.
35. Kaburaki, K., Mozumi, M., & Hasegawa, H., (2018). Estimation of two-dimensional motion velocity using ultrasonic signals beamformed in Cartesian coordinate for measurement of cardiac dynamics. *Japanese Journal of Applied Physics*, *57*(7S1), 07LF03.
36. Kanoulas, E., Aslam, J. A., Sharp, G. C., Berbeco, R. I., Nishioka, S., Shirato, H., & Jiang, S. B., (2007). Derivation of the tumor position from external respiratory surrogates with periodical updating of the internal/external correlation. *Physics in Medicine & Biology*, *52*(17), 5443.
37. Karim, I. A., Mohammed, K., Aziz, N. F. A. A., & Hua, L. T., (2015). Comparative safety performance evaluation of profiled deck composite slab from the use of slope-intercept and partial shear methods. *World Acad. Sci. Eng. Technol. Int. J. Civil, Environ. Struct. Constr. Archit. Eng.*, *9*, 1047–53.
38. Kazinski, P. O., & Korolev, P. S., (2023). Multichannel scattering for the Schr\"{o} dinger equation on a line with different thresholds at both infinities. *arXiv preprint arXiv:2307.00473*.
39. Kirmaci, U. S., & Özdemir, M. E., (2004). On some inequalities for differentiable mappings and applications to special means of

real numbers and to midpoint formula. *Applied Mathematics and Computation*, *153*(2), 361–368.

40. Kirmaci, U. S., (2004). Inequalities for differentiable mappings and applications to special means of real numbers and to midpoint formula. *Applied Mathematics and Computation*, *147*(1), 137–146.
41. Lee, H. J., Hong, Y., & Choi, H., (2017). Perceptions of tutoring roles and psychological distance among instructors, tutors, and students at a Korean university. *Higher Education Research & Development*, *36*(1), 143–157.
42. Li, X., Blackman, C. J., Choat, B., Duursma, R. A., Rymer, P. D., Medlyn, B. E., & Tissue, D. T., (2018). Tree hydraulic traits are coordinated and strongly linked to climate-of-origin across a rainfall gradient. *Plant, Cell, & Environment*, *41*(3), 646–660.
43. Long, A. W., Wolfe, K. C., Mashner, M. J., & Chirikjian, G. S., (2013). The banana distribution is Gaussian: A localization study with exponential coordinates. *Robotics: Science and Systems VIII*, *265*, 1.
44. Monari Martinez, E., & Neodo, K., (2022). Working with the Cartesian plane and algebraic formulas by students with down syndrome: Findings of a small-scale study. *International Journal of Disability, Development, and Education*, *69*(2), 609–622.
45. Nemirovsky, R., Rasmussen, C., Sweeney, G., & Wawro, M., (2012). When the classroom floor becomes the complex plane: Addition and multiplication as ways of bodily navigation. *Journal of the Learning Sciences*, *21*(2), 287–323.
46. Nguyen, V., Martinelli, A., Tomatis, N., & Siegwart, R., (2005). A comparison of line extraction algorithms using 2D laser rangefinder for indoor mobile robotics. In: *2005 IEEE/RSJ International Conference on Intelligent Robots and Systems* (pp. 1929–1934). IEEE.
47. Niehrs, C., (2010). On growth and form: a Cartesian coordinate system of Wnt and BMP signaling specifies bilaterian body axes. *Development*, *137*(6), 845–857.
48. Nishikawa, H., (2020). A face-area-weighted ‘centroid’ formula for finite-volume method that improves skewness and convergence on triangular grids. *Journal of Computational Physics*, *401*, 109001.
49. Oszczak, B., & Sitnik, E., (2013). The algorithm for determining the coordinates of a point in three-dimensional space by using the auxiliary point. *Artificial Satellites*, *48*(4), 141–145.

50. Ozcelik, E., & Acarturk, C., (2011). Reducing the spatial distance between printed and online information sources by means of mobile technology enhances learning: Using 2D barcodes. *Computers & Education*, *57*(3), 2077–2085.
51. Petzold, B., Reiss, P., & Stössel, W., (1999). Laser scanning—surveying and mapping agencies are using a new technique for the derivation of digital terrain models. *ISPRS Journal of Photogrammetry and remote Sensing*, *54*(2–3), 95–104.
52. Ray, N., (2005). PATHMATRIX: A geographical information system tool to compute effective distances among samples. *Molecular Ecology Notes*, *5*(1), 177–180.
53. Ross, N. P., DaAngela, J., Shanks, T., Wake, D. A., Cannon, R. D., Edge, A. C., & Weilbacher, P. M., (2007). The 2dF-SDSS LRG and QSO survey: the LRG 2-point correlation function and redshift-space distortions. *Monthly Notices of the Royal Astronomical Society*, *381*(2), 573–588.
54. Rote, G., (1991). Computing the minimum Hausdorff distance between two-point sets on a line under translation. *Information Processing Letters*, *38*(3), 123–127.
55. Ryckaert, J. P., Ciccotti, G., & Berendsen, H. J., (1977). Numerical integration of the cartesian equations of motion of a system with constraints: molecular dynamics of n-alkanes. *Journal of Computational Physics*, *23*(3), 327–341.
56. Sablin, M. N., (2020). Difference differential operators with values at the midpoints of the sides of cells of a triangular grid. *Moscow University Computational Mathematics and Cybernetics*, *44*, 97–108.
57. Saha, R. A., Ayub, A. F. M., & Tarmizi, R. A., (2010). The effects of GeoGebra on mathematics achievement: Enlightening coordinate geometry learning. *Procedia-Social and Behavioral Sciences*, *8*, 686–693.
58. Setiawan, Y. E., (2022). Analysis of error in determining the distance between two points in the cartesian plane in the first semester students. *Jurnal Pendidikan Matematika dan IPA*, *13*(1), 25–39.
59. Setzmann, U., & Wagner, W., (1991). A new equation of state and tables of thermodynamic properties for methane covering the range from the melting line to 625 K at pressures up to 1,000 MPa. *Journal of Physical and Chemical Reference Data*, *20*(6), 1061–1155.

60. Sterling, T. D., Perry, H., & Katz, L., (1964). Automation of radiation treatment planning—IV. Derivation of a mathematical expression for the per cent depth dose surface of cobalt 60 beams and visualization of multiple field dose distributions. *The British Journal of Radiology*, *37*(439), 544–550.
61. Stuart, J. T., (1954). On the stability of viscous flow between parallel planes in the presence of a co-planar magnetic field. *Proceedings of the Royal Society of London. Series A: Mathematical and Physical Sciences*, *221*(1145), 189–206.
62. Tamura, H., & Tanno, K., (2008). Midpoint validation method for support vector machine with margin adjustment technique. In: *2008 3rd International Conference on Innovative Computing Information and Control* (pp. 492–492). IEEE.
63. Thompson, H. B., (1967). Calculation of cartesian coordinates and their derivatives from internal molecular coordinates. *The Journal of Chemical Physics*, *47*(9), 3407–3410.
64. Timmons, J. H., Rudolf, P., Martell, A. E., Martin, J. W., & Clearfield, A., (1980). Crystal and molecular structures of complexes of two isomeric 18-membered tetraaza macrocyclic ligands having the empirical formula $[CuC_{20}H_{40}N_4](ClO_4)_2$. Effects of chelate ring size and double-bond placement on coordination geometry about copper (II). *Inorganic Chemistry*, *19*(8), 2331–2338.
65. Tsai, R., (1987). A versatile camera calibration technique for high-accuracy 3D machine vision metrology using off-the-shelf TV cameras and lenses. *IEEE Journal on Robotics and Automation*, *3*(4), 323–344.
66. Umek, W. H., Morgan, D. M., Ashton-Miller, J. A., & DeLancey, J. O., (2004). Quantitative analysis of uterosacral ligament origin and insertion points by magnetic resonance imaging. *Obstetrics and Gynecology*, *103*(3), 447.
67. Urankar, L., (1990). Common compact analytical formulas for computation of geometry integrals on a basic cartesian sub-domain in boundary and volume integral methods. *Engineering Analysis with Boundary Elements*, *7*(3), 124–129.
68. Van Calster, H., Baeten, L., De Schrijver, A., De Keersmaeker, L., Rogister, J. E., Verheyen, K., & Hermy, M., (2007). Management driven changes (1967–2005) in soil acidity and the understorey plant community following conversion of a coppice-with-standards forest. *Forest Ecology and Management*, *241*(1–3), 258–271.

69. Volodymyr, K., & Anatolii, M., (2018). The problem of forms of completing the Copernicus revolution in modern Cartesian science. Философия и космология, *21*, 131–139.

70. Voronina, E. P., (2021). Comprehensive socio-economic development of Arctic territories of the Russian Federation: Case of risks in the coordinate plane. In: *IOP Conference Series: Earth and Environmental Science* (Vol. 625, No. 1, p. 012012). IOP Publishing.

71. Walker, B. J., Skabelund, D. C., Busch, F. A., & Ort, D. R., (2016). *An Improved Approach for Measuring the Impact of Multiple CO_2 Conductance on the Apparent Photorespiratory CO_2 Compensation Point Through Slope–Intercept Regression* (Vol. 39, No. 6, pp. 1198–1203).

72. Wesolowsky, G. O., & Love, R. F., (1971). Location of facilities with rectangular distances among point and area destinations. *Naval Research Logistics Quarterly*, *18*(1), 83–90.

73. Yang, L., Powell, D. R., & Houser, R. P., (2007). Structural variation in copper (I) complexes with pyridylmethylamide ligands: structural analysis with a new four-coordinate geometry index, τ 4. *Dalton Transactions*, *9*, 955–964.

74. Yang, Z. F., & Tsai, W. H., (1998). Using parallel line information for vision-based landmark location estimation and an application to automatic helicopter landing. *Robotics and Computer-Integrated Manufacturing*, *14*(4), 297–306.

75. Yu, H., Huang, Y., Zheng, D., Bai, L., & Han, J., (2020). Three-dimensional shape measurement technique for large-scale objects based on line structured light combined with industrial robot. *Optik*, *202*, 163656.

76. Yuxiang, L., & Souplet, P., (2010). Single-point gradient blow-up on the boundary for diffusive Hamilton-Jacobi equations in planar domains. *Communications in Mathematical Physics*, *293*(2), 499–517.

77. Zhang, D., He, F., Han, S., Zou, L., Wu, Y., & Chen, Y., (2017). An efficient approach to directly compute the exact Hausdorff distance for 3D point sets. *Integrated Computer-Aided Engineering*, *24*(3), 261–277.

78. Zou, D., Chen, K., Rao, Z., Cao, J., & Liao, W. H., (2022). Design of a quad-stable piezoelectric energy harvester capable of programming the coordinates of equilibrium points. *Nonlinear Dynamics*, *108*(2), 857–871.

CHAPTER 5

Introduction to Similarity

Contents

5.1. INTRODUCTION

The concept of similarity holds a pivotal role in geometry. Similarity refers to the property of two or more geometric figures with the same shape but may differ in size. The chapter explores the intricacies of similarity, its applications, and the geometrical framework underlying This fundamental geometric principle (Kellenberger et al., 2011).

Similarity is the correspondence between two or more shapes or figures where the angles between corresponding sides are equal, and the ratio of the lengths of corresponding sides is constant. The constancy of ratios distinguishes similar figures from merely having the same shape (Son, 2013).

Mathematically, if we have two figures, say Figure A and Figure B, we can denote similarity as follows:

Figure A ~ Figure B

This symbol (~) represents similarity, meaning that Figure A is similar to Figure B.

5.1.1. Understanding the Concept of 'Similar Figures'

To comprehend the concept of similar figures, consider two triangles, triangle ABC and triangle DEF, as shown in Figure 5.1 (Restrepo-Ochoa et al., 2023).

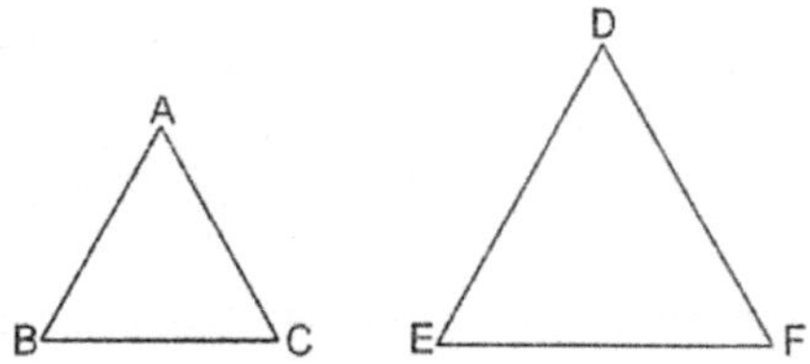

Figure 5.1. Triangle ABC and DEF are similar.

Source: Toppr, Creative Commons License.

Although triangle ABC and triangle DEF have different sizes in this illustration, their shapes remain the same. This fundamental idea is the essence of similarity – figures with identical shapes but possibly distinct sizes (Arican & Özçakir, 2021).

Several key characteristics define similar shapes (Park et al., 2010):

1. **Equal Corresponding Angles:** In similar figures, corresponding

angles are equal. In the case of Triangle ABC and Triangle DEF, $(\square A \equiv \square D), (\square B \equiv \square E), and (\square C \equiv \square F)$.

2. **Constant Ratio of Corresponding Sides:** The lengths of corresponding sides in similar figures are constant. In this case, $(AB / DE = BC / EF = AC / DF)$.
3. **Same Shape:** Similar figures possess the same shape. In our example, both triangles share the shape of a right triangle.

5.1.2. Similarity in Solving Real-World Problems

Similarity plays a crucial role in solving various real-world problems. One prominent application is in map scaling and cartography. Maps represent real geographical areas reduced to fit onto paper or screens. These scaled-down versions maintain the relative positions of landmarks and geographic features, relying on the principle of similarity (Modestou & Gagatsis, 2007).

For instance, consider creating a map of a sprawling city. Cartographers employ scaling to fit the city's roads, buildings, and landmarks on a standard-sized map. The map is a smaller, scaled city version, ensuring that distances and angles between landmarks remain accurate. This is possible due to the similarity between the city's layout and the map, allowing for precise navigation and planning (Lee & Yim, 2014). Another practical application of similarity is in the estimation of heights. Imagine measuring the height of a tall tree, a skyscraper, or a mountain. Often, it is not feasible to access these heights directly. Here, we can employ similarity principles, utilizing smaller objects whose heights are accessible for measurement. By calculating the scale factor, we can determine the height of the larger object.

5.1.3. The Concept of 'Scale Factor'

The concept of a scale factor is fundamental in understanding similarity. It represents the constant ratio by which corresponding lengths in similar figures are multiplied to transform one figure into another. Mathematically, if we denote the scale factor as 'k,' it can be expressed as (Cardone et al., 2003):

$$k = \frac{\text{Length of Corresponding Side in Figure A}}{\text{Length of Corresponding Side in Figure B}}$$

The scale factor 'k' allows us to transition from one figure to another while preserving their similarity. For instance, if 'k' equals 2, then every side in Figure A is twice as long as the corresponding side in Figure B.

The scale factor 'k' also aids in quantifying the enlargement or reduction of figures. An 'enlargement' occurs when 'k' is greater than 1, increasing in size, while a 'reduction' occurs when 'k' is less than 1, causing a decrease in size (Weng et al., 2019).

To illustrate, consider two rectangles, Rectangle P and Rectangle Q. If the length of the corresponding side of Rectangle P is double that of Rectangle Q, the scale factor 'k' is 2, and Its enlargement results in a larger Rectangle P compared to Rectangle Q.

5.2. SIMILARITY TRANSFORMATIONS

Similarity transformations are fundamental operations that involve changing the size and position of geometric figures while preserving their shape. This section explores one of the most prominent similarity transformations: dilations. We will study the definition and properties of dilations, understand how to find the scale factor of dilations, examine the relationship between dilations and similar figures, and provide practical examples illustrating dilations and their real-world applications (Weng et al., 2019).

5.2.1. Dilations

Dilation is a geometric transformation that enlarges or reduces a figure by a fixed scale factor while maintaining its shape. It essentially stretches or shrinks the figure uniformly in all directions from a fixed center called the "center of dilation."

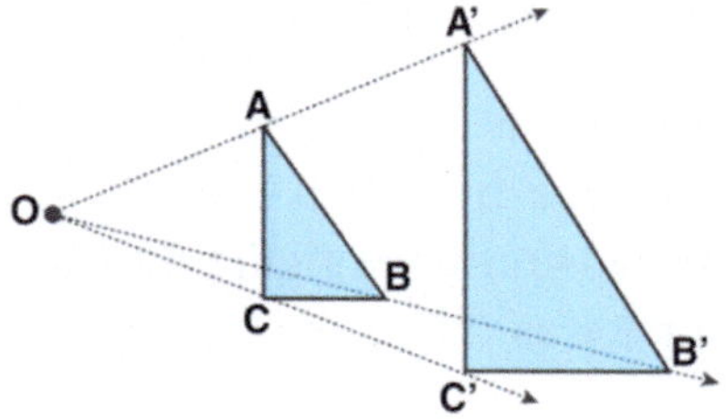

Figure 5.2. Dilation of a triangle ABC.

Source: BYJU's, Creative Commons License.

Mathematically, if we have a figure 'ABC' centered at point ')' and we want to dilate it by a scale factor 'k,' the resulting figure 'ABC' after dilation will have all its points 'k' times farther from the center 'O' compared to the original figure. It can be expressed as (Figure 5.2):

$$\text{Dilation}\left(ABC, O, k\right)$$

The following are the properties of dilations (De Bock et al., 1998):

1. **Preservation of Shape:** The most crucial property of dilations is that they preserve the shape of the figure. All angles in the dilated figure remain congruent to their corresponding angles in the original figure.
2. **Scaling Factor:** The scale factor 'k' determines the amount of enlargement or reduction. If 'k' is greater than 1, the figure is enlarged; if 'k' is between 0 and 1, it is reduced; if 'k' is 1, there is no change in size.
3. **Center Invariance:** The center of dilation remains unchanged after the transformation. It is a fixed point in the plane.
4. **Parallel Lines:** Lines passing through the center of dilation and their corresponding lines in the original figure are parallel.

5.2.2. Scale Factor of Dilations

To find the scale factor 'k' of a dilation, you can use the following formula (Lin & Lin, 2014):

$$k = \frac{\text{Length of a segment in the dilated figure}}{\text{Length of the corresponding segment in the original figure}}$$

The formula calculates how much each segment in the figure has been stretched or shrunk during dilation.

5.2.3. Dilations and Similar Figures

Dilations are closely related to the concept of similar figures. When two figures are similar, they have the same shape but possibly different sizes. A dilation with a scale factor 'k' transforms one figure into another similar figure. In other words, dilations are a specific type of similarity transformation.

Tip

When dealing with similarity transformations, remember that dilations are a fundamental concept. They involve scaling an object up or down while maintaining its shape. The scale factor is crucial in understanding how much the object is resized.

Mathematically, if Figure A is similar to Figure B, we can express this relationship using the similarity symbol (~):

$$\text{Figure A} \sim \text{Figure B}$$

Now, if we perform a dilation on Figure A with a scale factor 'k,' we get Figure A' (the dilated figure), and it will also be similar to Figure B:

$$\text{Dilation}\left(A, C, k\right) \sim \text{Figure B}$$

It demonstrates how dilations are fundamental in creating and understanding similar figures.

Following are the examples illustrating dilations:

1. **Map Scaling:** Consider a city map where a 1-inch segment represents 1 mile. To create a larger map, you can perform a dilation with a scale factor of 2. It means every mile on the original map now corresponds to 2 miles on the enlarged map. It is a practical application of dilations in cartography.
2. **Architectural Blueprints:** Architects use dilations to create scaled-down or scaled-up versions of building blueprints. For example, a blueprint might be scaled down by a factor of 0.5, making it half the size of the original plan while preserving the proportions and layout.
3. **Photography:** Zooming in or out with a camera lens is a dilation operation. Zooming in is a dilation that enlarges the scene in the frame while zooming out is a dilation that reduces it.

5.2.4. Similarity Transformations in 2D and 3D

While we have primarily discussed dilations in two dimensions (2D), it is essential to note that similarity transformations also apply to three-dimensional (3D) geometry. In 3D, dilations involve uniformly stretching or compressing objects in all directions from a fixed center of dilation. This concept is integral to 3D modeling, computer graphics, and architectural design (Lin & Lin, 2014). In summary, similarity transformations, particularly dilations, are fundamental tools in geometry that allow us to change the size of figures while preserving their shape. Dilations have essential properties, and their scale factor 'k' determines the degree of enlargement or reduction. Understanding dilations is crucial when dealing with similar figures, and they find practical applications in various fields, including cartography, architecture, and photography. Additionally, similarity transformations extend into three-dimensional geometry, where they play a vital role in various applications.

5.3. PROPORTIONAL FIGURES AND RATIOS

In this section, we will study the concept of proportional figures and the fundamental role of ratios in establishing similarity between geometric shapes. We explore proportionality between corresponding sides and angles, discuss the significance of the similarity ratio, and understand how ratios are employed to determine if figures are similar. Additionally, we introduce the 'Angle-Angle (AA)' criterion, which is a vital tool for establishing similarity between figures (Shi et al., 2021).

5.3.1. Proportionality between Corresponding Sides

One of the key aspects of similar figures is that their corresponding sides are in proportion. It means that the ratios of the lengths of corresponding sides in similar figures are equal. If we have two similar figures, Figure A and Figure B, and 'a' is the length of a side in Figure A while 'b' is the length of the corresponding side in Figure B, the proportionality can be expressed as (Ruff et al., 1989):

$$\frac{a}{b} = \frac{\text{Length of another side in Figure A}}{\text{Length of the corresponding side in Figure B}}$$

This proportionality holds for all pairs of corresponding sides.

5.3.2. Proportionality between Corresponding Angles

Similar figures also exhibit proportionality between corresponding angles. It means that the measures of corresponding angles in similar figures are equal. If angle 'A' in Figure A corresponds to angle 'B' in Figure B, this can be expressed as:

$$\square A = \square B$$

This relationship holds for all pairs of corresponding angles in similar figures (Bush & Karp, 2013).

5.3.3. Understanding the Similarity Ratio

The concept of the similarity ratio is central to establishing similarity between figures. The similarity ratio is the ratio of corresponding lengths in similar figures. For similar figures, this ratio remains constant. If we have a pair of similar figures, Figure A and Figure B, the similarity ratio can be expressed as (Castro et al., 2022):

"Similarity"

$$\text{Ratio} = \frac{\text{Length of a side in Figure A}}{\text{Length of the corresponding side in Figure B}}$$

This ratio directly measures how much larger or smaller Figure A is compared to Figure B.

5.3.4. Using Ratios to Determine Similarity

Ratios play a pivotal role in determining whether two figures are similar. We check if the figures' corresponding sides' ratios are equal to establish similarity. If all ratios are equal, the figures are similar. For instance, if we have three pairs of corresponding sides with ratios (Fuson & Abrahamson, 2005):

$$\frac{a}{b} = \frac{x}{y} = \frac{p}{q}$$

It implies that Figures A, B, and another figure are similar.

5.3.5. 'Angle-Angle (AA)' Criterion in Similarity

The 'angle-angle (AA)' criterion is a crucial tool for proving the similarity of two triangles. According to the criterion, the two triangles are similar if two angles in one triangle are congruent to two angles in another. This criterion is derived from the fact that if two angles of a triangle are known, the third angle can be calculated because the sum of the angles in any triangle is always 180°.

Mathematically, if $(\Box A)$ in triangle ABC is congruent to $(\Box X)$ in triangle XYZ, and $(\Box B)$ in triangle ABC is congruent to $(\Box Y) in$ triangle XYZ, then we can conclude:

$$\triangle ABC \sim \triangle XYZ$$

The AA criterion simplifies the process of proving similarity between triangles, allowing us to focus solely on angle measurements (Raju et al., 2022).

Hence, proportionality between corresponding sides and angles, the concept of the similarity ratio, and the AA criterion are essential tools for establishing similarity between geometric figures. These concepts have wide-ranging applications in architecture, cartography, art, and engineering, where precise scaling and proportionality are crucial.

5.4. PROVING SIMILARITY

This section will explore various criteria for proving the similarity between geometric figures. These criteria provide a systematic approach to establishing similarity and are essential tools in geometric analysis.

5.4.1. Side-Side-Side (SSS) Criterion for Similarity

The Side-Side-Side (SSS) criterion is a powerful method for proving the similarity between two triangles. According to the criterion, if the ratios of corresponding sides in two triangles are equal, the triangles are similar. Mathematically, if we have two triangles, ΔABC and ΔXYZ, and the ratios of corresponding sides are equal shapes (Weng et al., 2019):

$$\frac{AB}{XY} = \frac{BC}{YZ} = \frac{CA}{ZX}$$

Then we can conclude:

ÄABC ~ *ÄXYZ*

The SSS criterion provides a straightforward way to establish similarity between triangles based solely on side lengths (Figure 5.3).

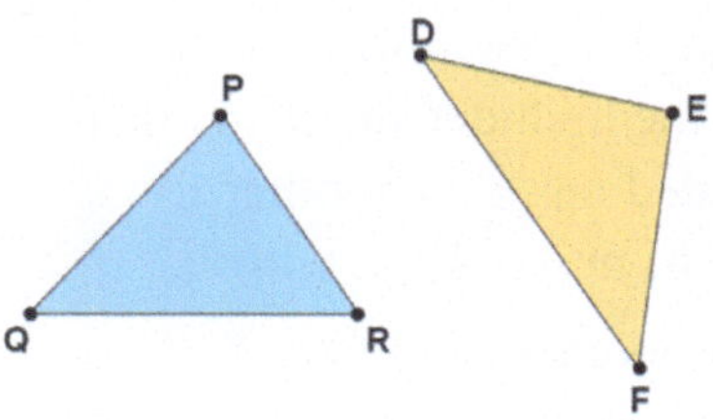

Figure 5.3. Illustration of SSS similarity criterion.

Source: CueMath, Creative Commons License.

5.4.2. Angle-Side-Angle (ASA) Criterion

The Angle-Side-Angle (ASA) criterion is another method for proving similarity between triangles. According to this criterion, if two angles and the included side in one triangle are congruent to two angles and the included side in another triangle, then the triangles are similar. Mathematically, if we have two triangles, ΔABC and ΔXYZ, with (Palinkas et al., 2015):

i. $\angle A \cong \angle X$;

ii. $\angle B \cong \angle Y$;

iii. Side AB ≅ Side XY.

Then we can conclude:

ÄABC ~ ÄXYZ

The ASA criterion is particularly useful when we have information about angle measures and the length of one side (Figure 5.4).

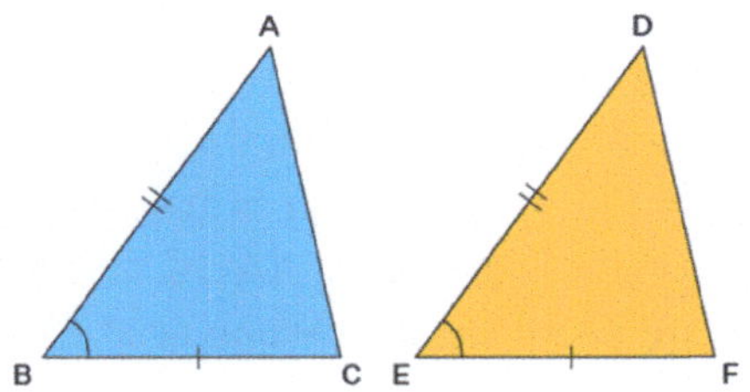

Figure 5.4. Schematic of ASA criterion.

Source: CueMath, Creative Commons License.

5.4.3. Side-Angle-Side (SAS) Criterion

The Side-Angle-Side (SAS) criterion is another approach for proving similarity between triangles. According to the criterion, the triangles are similar if two sides and the included angle in one triangle are proportional to two sides and the included angle in another triangle. Mathematically, if we have two triangles, ΔABC and ΔXYZ, with:

i. Side AB is proportional to side XY;

ii. Side BC is proportional to side YZ;

iii. $\angle B \cong \angle Y$.

Then we can conclude:

ÄABC ~ *ÄXYZ*

The SAS criterion combines side lengths and angle measures to establish similarity (Fisher et al., 2006).

5.4.4. Hypotenuse-Leg (HL) Criterion in Right Triangles

In right triangles, we have a specialized criterion known as the Hypotenuse-Leg (HL) criterion. This criterion applies specifically to right triangles and simplifies the process of proving their similarity. According to the HL criterion, if the hypotenuse and one leg of one right triangle are proportional to the hypotenuse and one leg of another right triangle, then the triangles are similar.

Mathematically, if we have two right triangles, ΔABC and ΔXYZ, with:

i. Hypotenuse AB is proportional to hypotenuse XY; and

ii. Leg BC is proportional to leg YZ.

Then we can conclude:

ÄABC ~ *ÄXYZ*

The HL criterion is valuable for demonstrating similarity in right triangles without comparing all sides and angles.

Let's consider a practical example problem to illustrate the application of these similarity criteria (Schuller et al., 2009):

Ø **Example:** Given two triangles, ΔABC and ΔXYZ, with the following side lengths and angle measures:

AB = 4 cm, XY = 6 cm.

BC = 5 cm, YZ = 7.5 cm.

CA = 6 cm, ZX = 9 cm.

$\angle A \cong \angle X$

$\angle B \cong \angle Y$

Using the criteria discussed, determine whether ΔABC is similar to ΔXYZ.

- **Solution:**

1. Using the SSS criterion, we compare the ratios of the corresponding sides:

$$\frac{AB}{XY} = \frac{4}{6} = \frac{2}{3}$$

$$\frac{BC}{YZ} = \frac{5}{7.5} = \frac{2}{3}$$

$$\frac{CA}{ZX} = \frac{6}{9} = \frac{2}{3}$$

Since all three ratios are equal, we can conclude that ΔABC is similar to ΔXYZ based on the SSS criterion.

2. Using the ASA criterion, we note that ∠A ≅ ∠X and ∠B ≅ ∠Y. Additionally, side AB ≅ side XY. Therefore, we can conclude that ΔABC is similar to ΔXYZ based on the ASA criterion.
3. Using the SAS criterion, we have side AB ≅ side XY, side BC ≅ side YZ, and ∠B ≅ ∠Y. Thus, we can conclude that ΔABC is similar to ΔXYZ based on the SAS criterion.

In this example, we applied multiple similarity criteria, leading to the same conclusion: ΔABC is similar to ΔXYZ.

5.4.5. Applications of Similarity and Proportionality

Let's consider a few real-world scenarios that demonstrate the significance of similarity and proportionality:

Remember

Similarity and proportionality concepts find practical use in a wide range of fields, from architecture to engineering. For instance, architects use similarity principles to create scaled models of buildings for design and presentation purposes.

- Ø **Example 1: Aerial Mapping:** It involves capturing images of large areas from aircraft or satellites. By analyzing these images and applying similarity principles, cartographers can create accurate maps for navigation, urban planning, and environmental studies (Schuller et al., 2009).
- Ø **Example 2: Architectural Modeling:** Architects use scale models to represent buildings and structures. These models are typically created on a smaller scale but maintain the proportionality and similarity of the original design. It allows architects to visualize and plan constructions effectively (Yang et al., 2019).

- Ø **Example 3: Medical Imaging:** In medical imaging, magnetic resonance imaging (MRI) and computed tomography (CT) scans rely on similarity to create detailed images of the human body's internal structures. Understanding the proportions and similarities within the body is vital for accurate diagnoses (Jain et al., 2008).
- Ø **Example 4: Art Restoration:** When restoring artwork, conservators must replicate missing or damaged parts of a painting or sculpture. Similarity ensures that the restored sections match the original in proportions, scale, and style (Izadi et al., 2023).

In each scenario, similarity and proportionality are essential tools for solving complex problems and accurately representing the real world. Whether in cartography, architecture, medical imaging, or art restoration, the principles of similarity provide a mathematical foundation for addressing practical challenges and achieving reliable results.

5.5. THE GOLDEN RATIO AND ITS SIGNIFICANCE

The Golden Ratio, often represented by the Greek letter Phi (φ), is a mathematical concept that has fascinated mathematicians, artists, architects, and scientists for centuries.The Golden Ratio is an irrational number approximately equal to 1.618033988749895. It is defined such that if you have two quantities, "a" and "b," the ratio of their sum to the larger of the two quantities is equal to the ratio of the larger quantity to the smaller one:

$$j = (a+b)/a = a/b$$

In other words, when you divide a line into two parts such that the ratio of the whole line to the longer part is the same as the ratio of the longer part to the shorter part, you have created a line segment in the Golden Ratio (Beckmann & Izsák, 2015).

5.5.1. Golden Ratio and Fibonacci Numbers

The Golden Ratio is linked with the Fibonacci sequence0/, a series of numbers where each number is the sum of the two preceding ones (e.g., 0, 1, 1, 2, 3, 5, 8, 13, …). As you progress through the Fibonacci sequence, dividing each number by its predecessor approximates the Golden Ratio. For example:

5/3 ≈ 1.6667 (approximation of φ)

$8/5 \approx 1.6$ (another approximation of φ)

$13/8 \approx 1.625$ (yet another approximation of φ)

This relationship between the Golden Ratio and the Fibonacci sequencea showcases the pervasive nature of φ in mathematics.

5.5.2. Golden Ratio in Art, Architecture, and Nature

In art, Golden Ratio is often associated with aesthetic harmony and balance. Many famous paintings, such as Leonardo da Vinci's "Mona Lisa" and Salvador Dalí's "The Sacrament of the Last Supper," are believed to incorporate the Golden Ratio in their compositions (Jacobson & Izsák, 2014).

Did You Know?

Golden ratio has been used in the design of famous structures like the Parthenon and is found in various natural phenomena, such as the spirals in seashells and the branching of trees.

In architecture, iconic structures like the Parthenon in Athens and the Pyramids of Egypt exhibit proportions influenced by the Golden Ratio. Architects have long appreciated how φ can create visually pleasing designs (Figure 5.5).

Figure 5.5. Employment of the Phi in the great pyramid of Giza.

Source: Illustrarch, Creative Commons License.

Nature is replete with examples of the Golden Ratio. In botany, the spirals of pinecones, sunflower seeds, and pineapples often follow Fibonacci patterns, demonstrating the inherent presence of the Golden Ratio in plant growth. Even the arrangement of leaves on some plants forms spirals in the Golden Ratio.

5.5.3. Golden Ratio in Geometric Constructions

The Golden Ratio can be calculated using its definition: $\varphi = (1 + \sqrt{5})/2$. This formula allows mathematicians and artists to create geometric constructions that adhere to the Golden Ratio. For example, a golden rectangle is one in which the ratio of the longer side to the shorter side is φ. The rectangle can be divided so.

To create a golden rectangle from a square, you extend one side to form a longer rectangle, maintaining the φ ratio. It can be repeated indefinitely, producing a series of nested golden rectangles, a design often seen in art and architecture (Jacobson & Izsák, 2014) (Figure 5.6).

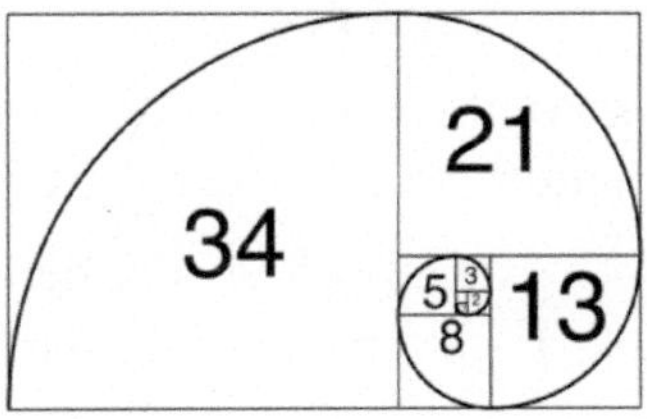

Figure 5.6. Golden rectangle with the first few terms from the Fibonacci sequence. Each number indicates the dimensions of the square it is in.

Source: StudyPug, Creative Commons License.

- Ø **Examples:** Here are some examples of how the golden ratio is used:
- • **Art: The Golden Rectangle.**
- o **Problem:** An artist wants to create a visually appealing composition using a rectangle that follows the Golden Ratio.
- o **Mathematical Equation (Golden Rectangle):** A Golden Rectangle has sides in the ratio of $/$:1. If you denote one side as "a," the other side will be "aj."

$$aj = a + a$$

This equation ensures that the ratio of the longer side (aj) to the shorter side (a) equals to $/$.

- **Solution:** The artist creates a rectangle with sides in this ratio, resulting in a visually pleasing composition.

- **Architecture: Parthenon's Facade**

  - **Problem:** Architects designing a building's facade want to incorporate the Golden Ratio into its proportions.

In the Parthenon, the ratio of the width of the facade to the height of the columns follows the Golden Ratio. If the width is "W" and the height of the columns is "H," then:

$$(W / H) = j$$

This equation ensures that the ratio of the building's width to the height of the columns is approximately equal to φ.

- **Solution:** Architects design the facade of the building, ensuring that the width-to-height ratio aligns with the Golden Ratio.

- **Nature: Fibonacci Sequence**

  - **Problem:** Observing the arrangement of leaves on a plant, you notice a spiral pattern that follows the Golden Ratio.

The Fibonacci Sequence is a series of numbers where each is the sum of the two preceding ones, typically starting with 0 and 1. The ratio of consecutive Fibonacci numbers approaches φ as the sequence progresses (Oktavianingtyas et al., 2018).

$$j \approx F(n+1) / F(n)$$

In this equation, "n" represents the position in the Fibonacci Sequence (Figure 5.7).

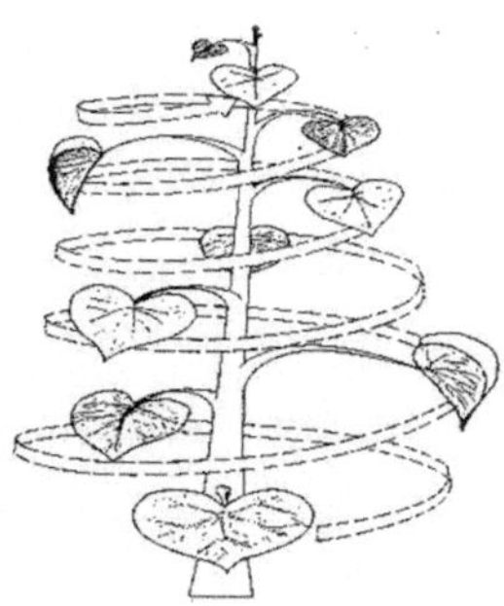

Figure 5.7. The Fibonacci series in the distribution of leaves around a central stem: 3 leaves in 5 turns, 5 leaves in 8 turns.

Source: Mehrdad Hejazi, Creative Commons License.

- o **Solution:** The spiral pattern observed in nature, such as in the arrangement of leaves on a stem or the seeds in a sunflower, follows the Fibonacci Sequence, and the ratio of consecutive Fibonacci numbers approaches the Golden Ratio.

These examples illustrate how the Golden Ratio, represented by the mathematical equation φ, is applied in art, architecture, and nature to create aesthetically pleasing proportions and patterns.

Activity 5.1

Imagine you are an architect tasked with designing a new skyscraper. How would you apply the principles of similarity and proportionality, including the use of the Golden Ratio, in your design to create an aesthetically pleasing and structurally sound building?

5.6. SUMMARY

This chapter straightforwardly introduces the concept of similarity. It begins by explaining what similar figures are and how they are essential in solving real-world problems. The chapter dives into similarity transformations, focusing on dilations and their scale factors. It discusses proportional figures, emphasizing the importance of corresponding sides and angles. Proving similarity using criteria like Side-Side-Side (SSS), Angle-Side-Angle (ASA), and others is explained. Lastly, the chapter touches upon the Golden Ratio (Phi), its link with Fibonacci numbers, and its widespread presence in art, architecture, and nature. Overall, this chapter provides a practical foundation for understanding and applying similarity concepts in various contexts.

REFERENCES

1. Arican, M., & Özçakir, B., (2021). Facilitating the development of preservice teachers' proportional reasoning in geometric similarity problems using augmented reality activities. *Education and Information Technologies*, *26*(2), 2327–2353.
2. Beckmann, S., & Izsák, A., (2015). Two perspectives on proportional relationships: Extending complementary origins of multiplication in terms of quantities. *Journal for Research in Mathematics Education*, *46*(1), 17–38.
3. Boto, E., Holmes, N., Leggett, J., Roberts, G., Shah, V., Meyer, S. S., & Brookes, M. J., (2018). Moving magnetoencephalography towards real-world applications with a wearable system. *Nature*, *555*(7698), 657–661.
4. Bush, S. B., & Karp, K. S., (2013). Prerequisite algebra skills and associated misconceptions of middle grade students: A review. *The Journal of Mathematical Behavior*, *32*(3), 613–632.
5. Cardone, A., Gupta, S. K., & Karnik, M., (2003). A survey of shape similarity assessment algorithms for product design and manufacturing applications. *J. Comput. Inf. Sci. Eng.*, *3*(2), 109–118.
6. Castro, P., Gómez, P., & Cañadas, M. C., (2022). Trends in learning and teaching of geometry: The case of the geometry and its applications meeting. *International Electronic Journal of Mathematics Education*, *17*(4), 654–656.
7. Cripe, D. G., & Randall, D. A., (2001). Joint variations of temperature and water vapor over the midlatitude continents. *Geophysical Research Letters*, *28*(13), 2613–2616.
8. De Bock, D., Verschaffel, L., & Janssens, D., (1998). The predominance of the linear model in secondary school students' solutions of word problems involving length and area of similar plane figures. *Educational Studies in Mathematics*, *35*(1), 65–83.
9. Enslein, K., & Brainerd, B., (1953). A minimal noise preamplifier for proportional counters and similar applications. *Review of Scientific Instruments*, *24*(10), 916–919.
10. Eustis, S., & El-Sayed, M. A., (2006). Why gold nanoparticles are more precious than pretty gold: Noble metal surface plasmon resonance and its enhancement of the radiative and nonradiative properties of nanocrystals of different shapes. *Chemical Society Reviews*, *35*(3), 209–217.

11. Fisher, S., Grice, E. A., Vinton, R. M., Bessling, S. L., & McCallion, A. S., (2006). Conservation of RET regulatory function from human to zebrafish without sequence similarity. *Science, 312*(5771), 276–279.
12. Fuson, K. C., & Abrahamson, D., (2005). Understanding ratio and proportion as an example of the apprehending zone and conceptual-phase problem-solving models. *Handbook of Mathematical Cognition, 45*(34), 213–234.
13. Hao, H., Nakayamada, S., Yamagata, K., Ohkubo, N., Iwata, S., Inoue, Y., & Tanaka, Y., (2021). Conversion of T follicular helper cells to T follicular regulatory cells by interleukin-2 through transcriptional regulation in systemic lupus Erythematosus. *Arthritis & Rheumatology, 73*(1), 132–142.
14. Hilaga, M., Shinagawa, Y., Kohmura, T., & Kunii, T. L., (2001). Topology matching for fully automatic similarity estimation of 3D shapes. In: *Proceedings of the 28th Annual Conference on Computer Graphics and Interactive Techniques* (Vol. 3, pp. 203–212).
15. Irfan, M., & Nusantara, T., (2020). Students know the concept but are incorrect in solving the proportional problem: How does it happen? *International Journal of Science, Mathematics & Technology Learning, 27*(2), 987–990.
16. Izadi, M., Alshuraiaan, B., Sheremet, M., & Galal, A. M., (2023). Melting process of a phase change material using single and double elliptic charger applied in thermal storage for building application: A numerical study on effect of position and aspect ratio. *Journal of Energy Storage, 72*, 108482.
17. Jacobson, E., & Izsák, A., (2014). Using coordination classes to analyze preservice middle-grades teachers' difficulties in determining direct proportion relationships. *Research Trends in Mathematics Teacher Education, 40*, 47–65.
18. Jain, P. K., Huang, X., El-Sayed, I. H., & El-Sayed, M. A., (2008). Noble metals on the nanoscale: optical and photothermal properties and some applications in imaging, sensing, biology, and medicine. *Accounts of Chemical Research, 41*(12), 1578–1586.
19. Jarrold, C., & Citroën, R., (2013). Reevaluating key evidence for the development of rehearsal: phonological similarity effects in children are subject to proportional scaling artifacts. *Developmental Psychology, 49*(5), 837.

20. Kellenberger, E., Hofmann, A., & Quinn, R. J., (2011). Similar interactions of natural products with biosynthetic enzymes and therapeutic targets could explain why nature produces such a large proportion of existing drugs. *Natural Product Reports*, *28*(9), 1483–1492.
21. Kesim, M., & Ozarslan, Y., (2012). Augmented reality in education: Current technologies and the potential for education. *Procedia-Social and Behavioral Sciences*, *47*, 297–302.
22. Koyuncu, I., Akyuz, D., & Cakiroglu, E., (2015). Investigating plane geometry problem-solving strategies of prospective mathematics teachers in technology and paper-and-pencil environments. *International Journal of Science and Mathematics Education*, *13*, 837–862.
23. Lee, H. S., & Yim, J., (2014). Pursuing coherence among proportionality, linearity, and similarity: Two pathways from preservice teachers' geometric representations. *The Mathematics Enthusiast*, *11*(3), 541–554.
24. Lian, Z., Godil, A., Sun, X., & Xiao, J., (2013). CM-BOF: Visual similarity-based 3D shape retrieval using clock matching and bag-of-features. *Machine Vision and Applications*, *24*, 1685–1704.
25. Lin, J. J. H., & Lin, S. S., (2014). Cognitive load for configuration comprehension in computer-supported geometry problem solving: An eye movement perspective. *International Journal of Science and Mathematics Education*, *12*, 605–627.
26. Ma, C., Zhu, Z., Fang, Z., Li, Z., & Liu, L., (2021). Optimal proportion of similar materials and rockburst tendency of white sandstone. *Advances in Materials Science and Engineering*, *2021*, 1–17.
27. Modestou, M., & Gagatsis, A., (2007). Students' improper proportional reasoning: A result of the epistemological obstacle of "linearity." *Educational Psychology*, *27*(1), 75–92.
28. Mooldijk, A., & Sonneveld, W., (2010). Coherent education in mathematics and physics: The theme of proportionality in mathematics and physics. *Trend in Science and Mathematics Education (TiSME)*, *40*, 43–50.
29. Oktavianingtyas, E., Salama, F. S., Fatahillah, A., Monalisa, L. A., & Setiawan, T. B., (2018). Development 3D animated story as interactive learning media with lectora inspire and plotagon on direct and inverse

proportion subject. In: *Journal of Physics: Conference Series* (Vol. 1108, pp. 121–125). IOP Publishing.

30. Pajares, F., & Miller, M. D., (1997). Mathematics self-efficacy and mathematical problem solving: Implications of using different forms of assessment. *The Journal of Experimental Education*, *65*(3), 213–228.
31. Palinkas, L. A., Horwitz, S. M., Green, C. A., Wisdom, J. P., Duan, N., & Hoagwood, K., (2015). Purposeful sampling for qualitative data collection and analysis in mixed method implementation research. *Administration and Policy in Mental Health and Mental Health Services Research*, *42*, 533–544.
32. Park, J. S., Park, J. H., & Kwon, O. N., (2010). Characterizing the proportional reasoning of middle school students. *The SNU Journal Of Education Research*, *20*, 119–144.
33. Raju, C. S. K., Ahammad, N. A., Sajjan, K., Shah, N. A., Yook, S. J., & Kumar, M. D., (2022). Nonlinear movements of axisymmetric ternary hybrid nanofluids in a thermally radiated expanding or contracting permeable Darcy Walls with different shapes and densities: Simple linear regression. *International Communications in Heat and Mass Transfer*, *135*, 106110.
34. Restrepo-Ochoa, J. F., Gualdrón-Pinto, E., & Ávila-Ascanio, L. F., (2023). Improving the learning of geometric proportionality using van Hiele's model, mathematical visualization, and GeoGebra. *EURASIA Journal of Mathematics, Science, and Technology Education*, *19*(9), em2324.
35. Ribe, N. M., & Davaille, A., (2013). Dynamical similarity and density (non-) proportionality in experimental tectonics. *Tectonophysics*, *608*, 1371–1379.
36. Ruff, C. B., Walker, A., & Teaford, M. F., (1989). Body mass, sexual dimorphism and femoral proportions of Proconsul from Rusinga and Mfangano Islands, Kenya. *Journal of Human Evolution*, *18*(6), 515–536.
37. Schuller, B., Müller, R., Eyben, F., Gast, J., Hörnler, B., Wöllmer, M., & Konosu, H., (2009). Being bored? Recognizing natural interest by extensive audiovisual integration for real-life application. *Image and Vision Computing*, *27*(12), 1760–1774.
38. Schützeichel, R., (2019). Architectural proportion and appropriate urban design. Investigating the principle of scale in urban design

through the work of Theodor Fischer. In: *Proportions and Cognition in Architecture and Urban Design* (Vol. 3, pp. 47–63). Dietrich Reimer Verlag.

39. Seago, N., Jacobs, J., Driscoll, M., Matassa, M., & Callahan, M., (2013). Developing teachers' knowledge of a transformations-based approach to geometric similarity. *Mathematics Teacher Educator*, *2*(1), 74–85.
40. Selvarasah, S., Chao, S. H., Chen, C. L., Sridhar, S., Busnaina, A., Khademhosseini, A., & Dokmeci, M. R., (2008). A reusable high aspect ratio parylene-C shadow mask technology for diverse micropatterning applications. *Sensors and Actuators A: Physical*, *145*, 306–315.
41. Shahzad, A., Tian, F. B., Young, J., & Lai, J., (2018). Effects of hawkmoth-like flexibility on the aerodynamic performance of flapping wings with different shapes and aspect ratios. *Physics of Fluids*, *30*(9), 423–425.
42. Shi, Y., Ye, Y., Hu, N., Huang, X., & Wang, X., (2021). Experiments on material proportions for similar materials with high similarity ratio and low strength in multilayer shale deposits. *Applied Sciences*, *11*(20), 9620.
43. Son, J. W., (2013). How preservice teachers interpret and respond to student errors: ratio and proportion in similar rectangles. *Educational Studies in Mathematics*, *84*, 49–70.
44. Stormer, D. G., & Juanes, F., (2016). Effects of temperature and ration on the otolith-to-somatic size relationship in juvenile Chinook salmon (*Oncorhynchus tshawytscha*): A test of the direct proportionality assumption. *Marine and Freshwater Research*, *67*(7), 913–924.
45. Tarp, A., (2018). Mastering many by counting, re-counting, and double-counting before adding on-top and next-to. *Journal of Mathematics Education*, *11*(1), 103–117.
46. Weng, H., Wang, S., Lin, X., Li, Z., & Huang, J., (2019). A novel criterion applicable to transformer differential protection based on waveform sinusoidal similarity identification. *International Journal of Electrical Power & Energy Systems*, *105*, 305–314.
47. Yang, L., Zhou, Z., Song, J., & Chen, X., (2019). Anisotropic nanomaterials for shape-dependent physicochemical and biomedical applications. *Chemical Society Reviews*, *48*(19), 5140–5176.

CHAPTER 6

Three-Dimensional Geometry

Contents

6.1. INTRODUCTION

In geometry, the transition from two to three dimensions brings a profound shift in perspective. Three-dimensional geometry is a branch of mathematics that studies objects and spaces in three dimensions. This chapter is a gateway to the world of 3 dimensions, laying the foundation for understanding and working within three-dimensional space (Maxelon & Mancktelow, 2005).

Three-dimensional space, often called 3D space, is the conceptual realm where objects possess three dimensions: length, width, and height. Unlike two-dimensional space, which is confined to flat surfaces, three-dimensional space extends in all directions, allowing for a more complex representation of our physical world (Maxelon & Mancktelow, 2005).

6.1.1. Cartesian Coordinate System in 3D

To navigate and describe 3D space systematically, we employ the Cartesian coordinate system, an extension of the familiar 2D Cartesian plane. In this system, three mutually perpendicular axes intersect at a common point, known as the origin. The axes are labeled as the x-axis, y-axis, and z-axis, corresponding to the three dimensions. The positive direction along each axis is defined according to the right-hand rule, ensuring consistency in spatial orientation (Lavaste et al., 1992).

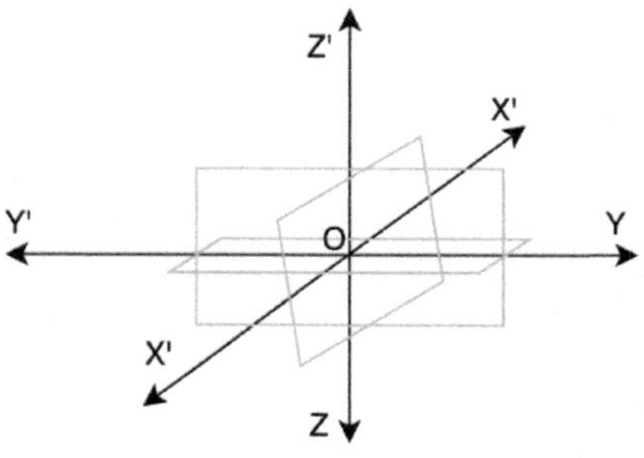

Figure 6.1. Cartesian coordinate system in 3D.

Source: Geeks for Geeks, Creative Commons License.

6.1.2. Coordinates of a Point in 3D

In the Cartesian coordinate system, any point in 3D space can be uniquely identified by an ordered triplet of numbers (x, y, z). These numbers represent the distances along each axis from the origin to the point. The x-coordinate specifies the position along the x-axis, the y-coordinate along the y-axis, and the z-coordinate along the z-axis. These coordinates precisely describe the point's location in 3D space (Darling & Gilchrist, 1991).

Mathematically, a point P in 3D space is denoted as:

$$P(x, y, z)$$

Here, (x, y, z) represents the coordinates of the point P.

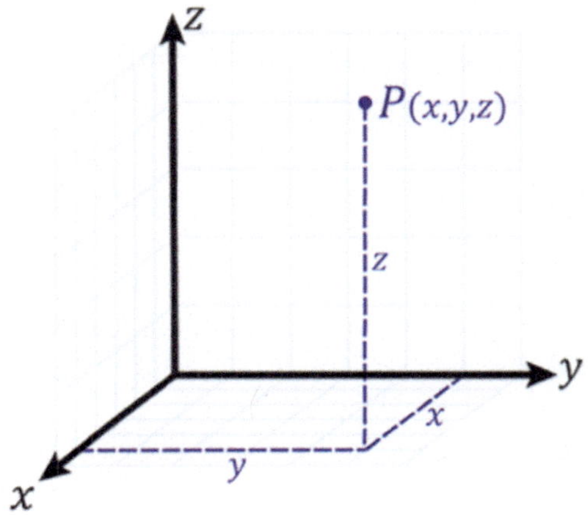

Figure 6.2. Illustration of a point P (x,y,z) in a 3D Cartesian coordinate system.

Source: Skill You Need, Creative Commons License.

6.1.3. Visualizing 3D Space

One of the initial challenges when studying three-dimensional geometry is visualizing 3D space. Unlike the intuitive and straightforward nature of 2D space, 3D space introduces complexity due to its depth. Visualizing objects and their relationships in 3D space requires a mental adaptation that extends beyond flat surfaces (Darling & Gilchrist, 1991).

Techniques such as perspective drawings, computer graphics, and physical models are commonly employed to facilitate the visualization. Perspective drawings use techniques like foreshortening to convey the illusion of depth on a two-dimensional surface. Computer graphics and modeling software provide interactive tools for creating and manipulating 3D objects, aiding in visualization and practical applications.

A fundamental concept that aids in visualizing 3D space is the notion of orthogonality. In 3D space, orthogonality signifies that two vectors or objects are perpendicular. This concept is crucial in three-dimensional geometry, from calculating angles between objects to determining the intersection of lines and planes (Darling & Gilchrist, 1991).

To illustrate the concept of orthogonality and the Cartesian coordinate system in 3D space, let's consider an example:

- Ø **Example: Orthogonal Vectors:** Suppose we have two vectors $(\vec{A})$ and $(\vec{B})$, in 3D space. Their components along the x, y, and z axes are given as follows:

$$\vec{A} = (2,1,3)$$

$$\vec{B} = (-1,4,-2)$$

We can use the dot product, a fundamental operation in 3D space, to determine whether these vectors are orthogonal. The dot product of two vectors $(\vec{A})$ and $(\vec{B})$ is given by:

$$\vec{A} \cdot \vec{B} = A_x \cdot B_x + A_y \cdot B_y + A_z \cdot B_z$$

Calculating the dot product:

$$\vec{A} \cdot \vec{B} = (2 \cdot -1) + (1 \cdot 4) + (3 \cdot -2) = -2 + 4 - 6 = -4$$

Since the dot product $(\vec{A} \cdot \vec{B})$ is not equal to zero, we conclude that vectors $(\vec{A})$ and $(\vec{B})$ are not orthogonal. This example demonstrates how vectors in 3D space can be analyzed for orthogonality, a concept critical for understanding spatial relationships and geometry in three dimensions (Liu et al., 2022).

6.2. DISTANCE AND MIDPOINT IN 3D SPACE

Understanding how to calculate distances between points and find midpoints is crucial in three-dimensional space. These concepts have practical applications in fields ranging from physics to computer graphics, making them fundamental in three-dimensional geometry (Dabaghi & Pironneau, 1986).

6.2.1. Calculating Euclidean Distance in 3D

The Euclidean distance between two points in 3D space measures the length of the shortest path connecting them. In mathematical terms, for two points, $P(x_1, y_1, z_1)$ and $Q(x_2, y_2, z_2)$, the Euclidean distance (d) can be computed using the following formula:

$$d = \sqrt{(x_2 - x_1)^2 + (y_2 - y_1)^2 + (z_2 - z_1)^2}$$

It employs the Pythagorean theorem in each dimension (x, y, and z) to calculate the distances separately and then combine them to find the overall distance. It is important to note that the square root is applied to obtain the actual distance (Xu et al., 2017).

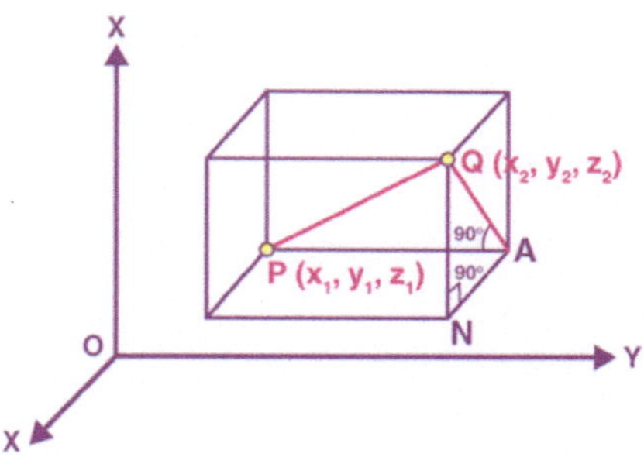

Figure 6.3. Illustration of distance in 3D space.

Source: BYJU's, Creative Commons License.

Ø **Example: Calculating 3D Distance**

Let's consider two points in 3D space:

Point A: $(1, 2, 3)$

Point B: $(4, 5, 6)$

Using the Euclidean distance formula:

$$d = \sqrt{(4-1)^2 + (5-2)^2 + (6-3)^2}$$

Calculating each component:

$$d = \sqrt{3^2 + 3^2 + 3^2}$$

$$d = \sqrt{9+9+9}$$

$$d = \sqrt{27}$$

The Euclidean distance (d) between points A and B is $(\sqrt{27})$ units.

6.2.2. Finding the Midpoint of Two Points in 3D

The midpoint of a line segment between two points in 3D space is the point that lies exactly halfway between them. To find the midpoint, you can average the coordinates of the two points. For points (x_1, y_1, z_1) and (x_2, y_2, z_2), the midpoint (x_m, y_m, z_m) is calculated as follows (Levit & Lasinski, 1991):

$$x_m = \frac{x_1 + x_2}{2}$$

$$y_m = \frac{y_1 + y_2}{2}$$

$$z_m = \frac{z_1 + z_2}{2}$$

This method takes the average of each coordinate to determine the midpoint in 3D space.

Ø **Example: Finding the Midpoint**

Let's use the points from the previous example, $A(1,2,3)$ and $B(4,5,6)$, to find the midpoint.

$$x_m = \frac{1+4}{2} = \frac{5}{2}$$

$$y_m = \frac{2+5}{2} = \frac{7}{2}$$

$$z_m = \frac{3+6}{2} = \frac{9}{2}$$

So, the midpoint $M\left(\frac{5}{2},\frac{7}{2},\frac{9}{2}\right)$ is exactly halfway between points A and B in 3D space.

In summary, the ability to calculate distances and find midpoints in three-dimensional space is essential for solving practical problems across various fields. These fundamental concepts strengthen our understanding of spatial relationships and enable precise measurements and calculations in 3D geometry (Sergeev et al., 1996).

6.3. EQUATIONS OF LINES AND PLANES IN 3D

In three-dimensional geometry, understanding how to represent lines and planes is crucial for solving complex spatial problems. Lines and planes are fundamental geometric elements, and in this section, we will study their equations and interactions (Kondolf et al., 2006).

Remember

Remember that in 3D geometry, finding the intersection of a line and a plane is similar to solving real-world problems. Just as a line intersects a surface, think of scenarios like an airplane's flight path intersecting with the Earth's surface, which is a fundamental concept in aviation and geography.

6.3.1. Parametric and Vector Equations of Lines

A line in three-dimensional space can be represented in two primary forms: parametric and vector equations.

1. **Parametric Equation:** For a line L, given by a point (x_0, y_0, z_0) and a direction vector (a,b,c), the parametric equations are:

$$x = x_0 + at$$

$$y = y_0 + bt$$

$$z = z_0 + ct$$

Here, (x, y, z) represents any point on the line, and (t) is a parameter that allows us to traverse the line.

2. **Vector Equation:** The vector equation of a line L is expressed as:

$$r = r_0 + vt$$

where; r is a position vector for any point on the line; r_0 is the position vector of a specific point on the line; v is the direction vector of the line; and is a parameter.

Ø **Example: Parametric and Vector Equations**

Consider a line L passing through the point (1, 2, 3) and parallel to the vector $2,-1,1$. Its parametric equations are:

$$x=1+2t$$

$$y=2-t$$

$$z=3+t$$

And the vector equation is:

$$r = 1,2,3 + t2,-1,1$$

6.3.2. Finding Intersection Points of Lines

To find the intersection point of two lines in 3D space, we must equal their respective parametric equations and solve for the parameter(s) at which they intersect (Amrouche et al., 1998). A plane in 3D space can be represented using two primary forms: general and vector equations.

1. **General Equation:** For a plane (Đ), given by a point (x_0, y_0, z_0) and a normal vector $n = a,b,c$, the general equation of the plane is:

$$ax+by+cz=d$$

where; $(d = ax_0 + by_0 + cz_0)$. This equation describes the plane's points (x, y, z).

2. **Vector Equation:** The vector equation of a plane (Đ) is given by:

$$n \cdot (r - r_0) = 0$$

where; n is the normal vector; r represents any point on the plane; and r_0 is the position vector of a specific point on the plane.

6.3.3. Intersection of a Line and a Plane

To find the intersection point of a line and a plane, we substitute the parametric equations of the line into the vector equation of the plane. It allows us to solve for the parameter at which the line intersects the plane (Amrouche et al., 1998).

Ø **Example: Intersection of Line and Plane**

Consider a plane (Đ) with a normal vector $n = 1, 2, -1$ and containing the point (2, –1, 3). Also, let's consider a line L defined by parametric equations:

$$x = 3 + t$$

$$y = -1 + 2t$$

$$z = 2t$$

To find the intersection point, we substitute these equations into the vector equation of the plane:

$$1, 2, -1 \cdot (3, -1, 2 + t1, 2, -1 - 2, -1, 3) = 0$$

Solving for (t) yields the parameter at which the line intersects the plane. Once we have (t), we can find the coordinates of the intersection point by substituting it into the parametric equations for the line.

Understanding the equations of lines and planes in 3D space and how they intersect is essential for solving spatial problems in fields like physics, engineering, computer graphics, and architecture. These equations provide a powerful toolset for modeling and analyzing three-dimensional complex geometric relationships (May, 1966).

6.4. VECTORS IN THREE-DIMENSIONAL SPACE

In three-dimensional geometry, vectors play a fundamental role in representing quantities with magnitude and direction. In this section, we will explore the representation of vectors in 3D, vector operations such as addition and subtraction, scalar multiplication, dot and cross products, and vector projections in three-dimensional space (Illés et al., 2011).

6.4.1. Representing Vectors in 3D

In three-dimensional space, a vector 'u' can be represented as an ordered triple x, y, z, where $(x), (y)$, and (z) represent the vector's components along the x, y, and z-axes, respectively. This representation allows us to visualize vectors as arrows originating from the origin and pointing to a specific point in 3D space (Ohtsuki & Tanaka, 2006).

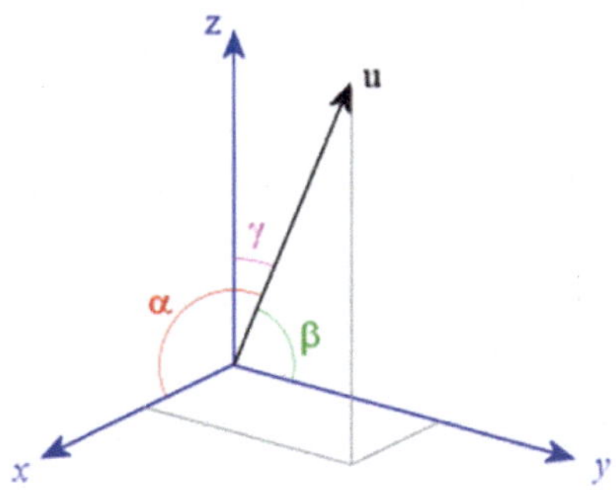

Figure 6.4. Representation of a vector ‘u’ in 3D space.

Source: Mathematics Stack Exchange, Creative Commons License.

1. **Vector Addition:** This in 3D involves adding the corresponding components of two vectors. If $A = A_x, A_y, A_z$ and $\left(B = B_x, B_y, B_z\right)$, then the sum $\left(A+B\right)$ is given by:

   $\vec{A}+\vec{B} = A_x + B_x, A_y + B_y, A_z + B_z$

In this equation:

i. $\left(\vec{A}\right)$ and $\left(\vec{B}\right)$ represent two three-dimensional vectors.

ii. $\left(A_x, A_y, A_z\right)$ are the components of the vector $\left(\vec{A}\right)$ in the x, y, and z directions, respectively.

iii. $\left(B_x, B_y, B_z\right)$ are the components of the vector $\left(\vec{B}\right)$ in the x, y, and z directions, respectively.

The notation uses standard vector notation with angle brackets (< >) to represent a vector in three-dimensional space.

2. **Subtraction of Vectors:** Subtracting one vector from another follows a similar pattern, where we subtract corresponding components. If $A = A_x, A_y, A_z$ and $B = B_x, B_y, B_z$, then the difference $(A-B)$ is given by:

   $A - B = A_x - B_x, A_y - B_y, A_z - B_z$

3. **Scalar Multiplication:** It involves multiplying a vector by a scalar (a real number). If $A = A_x, A_y, A_z$ and (k) is a scalar, then the scalar multiplication (kA) is given by:

$$kA = kA_x, kA_y, kA_z$$

4. **Dot Product:** The dot product of two vectors, denoted as $(A \cdot B)$, is a scalar quantity calculated as:

$$A \cdot B = A_x B_x + A_y B_y + A_z B_z$$

The dot product measures the projection of one vector onto another and is used in various applications, including work and angle calculations (Silvers, 1995).

5. **Cross Product:** The cross product of two vectors, denoted as $(A \times B)$, results in a vector that is perpendicular to both (A) and (B). Its components are calculated as:

$$A \times B = \begin{bmatrix} A_y B_z - A_z B_y, \\ A_z B_x - A_x B_z, \\ A_x B_y - A_y B_x, \end{bmatrix}$$

The cross-product is valuable in applications involving areas, volumes, and torque calculations.

6. **Vector Projections in 3D:** Vector projections allow us to find the component of one vector in the direction of another. For example, the projection of (A) onto (B) is given by:

$$\text{Proj}_{\text{B}}(\text{A}) = \frac{\text{A} \cdot \text{B}}{\|\text{B}\|} \frac{\text{B}}{\|\text{B}\|}$$

This projection provides a means to analyze vector quantities in a specific direction (Dolgopolov & Rodionova, 2011).

Ø **Example: Vector Operations in 3D:** Consider two vectors $A = 1, 2, -1 \, and \, B = 3, -1, 2$. We can perform various operations on these vectors (Fox & Artin, 1948):

1. Vector Addition:

$$A + B = 1 + 3, 2 - 1, -1 + 2 = 4, 1, 1$$

2. Vector Subtraction:

$$\mathrm{A} - \mathrm{B} = \langle 1-3, 2+1, -1-2 \rangle = \langle -2, 3, -3 \rangle$$

3. Scalar Multiplication:

$$2A = 2 \cdot 1, 2 \cdot 2, 2 \cdot (-1) = 2, 4, -2$$

4. Dot Product:

$$A \cdot B = (1 \cdot 3) + (2 \cdot (-1)) + (-1 \cdot 2) = 3 - 2 - 2 = -1$$

5. Cross Product:

$$A \times B = (2 \cdot 2 - (-1) \cdot (-1)), (-1 \cdot 3 - (-1) \cdot 1), (1 \cdot (-1) - 2 \cdot 3) = 3, -2, -5$$

Understanding vector operations is crucial in solving problems related to forces, motion, electromagnetic fields, and more in various fields, including physics, engineering, and computer graphics. Vectors provide a concise and powerful way to describe quantities with both magnitude and direction in three-dimensional space (Huang et al., 2021).

6.4.2. 3D Geometry: Surfaces and Solids

Three-dimensional geometry involves various types of surfaces and solids. We classify surfaces such as spheres, cylinders, cones, and more based on their defining characteristics.

For instance, a sphere is defined as the set of all points equidistant from a central point, while a cylinder has two parallel circular bases (Rathod & Rao, 1995. Understanding the surface area and volume of 3D solids is important in various fields, including architecture, engineering, and manufacturing.

1. **Surface Area:** The surface area of a 3D solid is the sum of the areas of its faces. For example, the surface area of a cylinder is given by the formula $(2pr^2 + 2prh)$, where (r) is the radius of the base and (h) is the height.
2. **Volume:** The volume of a 3D solid represents the space enclosed by its surfaces. For instance, the volume of a cone can be calculated using the formula $\left(\frac{1}{3}pr^2h\right)$, where (r) is the radius of the base and (h) is the height.

6.5. 3D COORDINATE SYSTEMS AND TRANSFORMATION MATRICES

This section will explore advanced topics in three-dimensional geometry, including alternate coordinate systems like cylindrical and spherical coordinates, transformations in 3D space, and their representation using matrices (Qiu et al., 2019).

6.5.1. Transformations in 3D Space

Understanding transformations in 3D space is significant in computer graphics, robotics, and engineering. We explore three fundamental transformations (Sastry et al., 1997):

1. **Translation:** This involves shifting an object's position in space. Mathematically, this transformation is represented as a vector $(\vec{T})$ describing the displacement in each coordinate $(x),(y),(z)$. For instance, if $\vec{T}=[3,1,2]$, it means moving 3 units along the x -axis, 1 unit along the y -axis, and 2 units along the z -axis.
2. **Rotation:** It changes the orientation of an object. We use rotation matrices or quaternions to describe these transformations. Euler angles or roll-pitch-yaw angles can also specify rotations. For example, a 90° rotation about the $z-$axis can be represented by a rotation matrix.
3. **Scaling:** This modifies the size of an object along each axis. It involves scaling factors $(S_x),(S_y),(S_z)$ that determine how much an object is stretched or compressed along the $(x),(y)$, and (z) directions. Scaling can be non-uniform, leading to different scaling factors along different axes.

6.5.2. Representation of Transformations Using Matrices

Transformation matrices provide a concise and efficient way to represent translations, rotations, and scaling in 3D space. These matrices allow us to combine multiple transformations into a single matrix for complex operations. For example, given the translation matrix (T) and rotation matrix (R), we can represent their composition as (TR) (Sastry et al., 1997).

Tip

When working with transformation matrices in 3D space, think of them as the instructions that guide objects through a 3D world. Just like in a video game, matrices can rotate, scale, or translate objects, making them essential in computer graphics and animation.

In conclusion, this chapter delves into advanced topics in three-dimensional geometry, including alternate coordinate systems like cylindrical and spherical coordinates, transformations in 3D space, and their representation using matrices. The practical applications of 3D transformations are far-reaching, encompassing fields as diverse as computer graphics, robotics, engineering, and medical imaging. Understanding these concepts is essential for professionals and students seeking to work with three-dimensional data and solve complex real-world problems (Sastry et al., 1997).

6.6. ANALYZING SPACE DIAGONALS AND ANGLES IN 3D SOLIDS

This section focuses on a deeper exploration of three-dimensional geometry by delving into space diagonals, their lengths, and the angles between them within various 3D solids, including tetrahedrons and cubes (Faugeras, 1995).

Did You Know?

that understanding space diagonals in 3D solids is crucial for structural engineers? By analyzing space diagonals, engineers can determine the stability and strength of complex 3D structures, from buildings to bridges.

6.6.1. Understanding Space Diagonals

In three-dimensional geometry, a space diagonal is a line segment that connects two non-adjacent vertices of a 3D solid. These diagonals traverse the interior of the solid and provide insight into its spatial properties.

6.6.1.1. Calculating Diagonal Lengths

Determining the lengths of space diagonals is essential for understanding the size and structure of 3D solids. The Pythagorean theorem is a fundamental tool for calculating diagonal lengths. For example, in a cube with side length

(a), the length of a space diagonal (d) can be calculated using the equation $d = \sqrt{a^2 + a^2 + a^2} = \sqrt{3}a.$

6.6.1.2. Angles Between Diagonals

Angles between space diagonals in 3D solids are crucial for understanding the spatial relationships within these shapes. In tetrahedrons, for instance, there are angles formed between various pairs of diagonals. These angles depend on the geometry of the solid and its symmetry (Soechting & Flanders, 1992).

6.6.2. Geometry of 3D Solids

In this section, we explore the space diagonals and angles within specific 3D solids:

1. **Tetrahedron:** It is a four-faced polyhedron with triangular faces. Investigating the angles between its diagonals helps us understand its inherent symmetry and geometric properties.
2. **Cube:** These are among the most well-known 3D solids, featuring six square faces. We calculate the lengths of space diagonals within cubes and explore the angles they form.
3. **Other 3D Solids:** The principles and techniques discussed in this chapter can be extended to other 3D solids, such as rectangular prisms, parallelepipeds, and more.

Understanding space diagonals and their angles is important for geometry enthusiasts and has practical applications in fields like engineering and architecture. For instance, knowing the angles between diagonals can aid in optimizing the design and stability of architectural structures.

6.7. APPLICATIONS OF THREE-DIMENSIONAL GEOMETRY

Some applications of 3D geometry are discussed in subsections.

6.7.1. Physics and Computer Science

In physics, 3D geometry is vital. It helps physicists' model and simulate physical phenomena in three-dimensional space. 3D geometry enables scientists to visualize and comprehend complex systems, from celestial mechanics to fluid dynamics. Moreover, computer science relies heavily on 3D geometry for computer graphics, computer-aided design (CAD), and animation. Algorithms for rendering 3D scenes and simulating 3D environments are built upon geometric principles (Pauwels et al., 2011).

Ø **Examples in Computer Graphics:**

1. **Perspective Projection:** The transformation from 3D coordinates (x, y, z) to 2D screen coordinates (X, Y) is given by:

$$X = f(x/z)$$

$$Y = f(y/z)$$

6.7.2. Medical Imaging and 3D Geometry

In medicine, 3D geometry is integral to advanced imaging techniques like magnetic resonance imaging (MRI) and computed tomography (CT) scans. These technologies generate detailed three-dimensional representations of the human body's internal structures. Surgeons use this data for preoperative planning and precise interventions, minimizing risks during procedures (Picos et al., 2016).

Ø **Example: 3D Reconstruction Algorithm**

The equation for three-dimensional reconstruction is a fundamental tool in medical imaging used to convert a series of two-dimensional slices into a comprehensive three-dimensional representation of an object or anatomical structure. It is expressed as follows:

$$V(x,y,z) = \iiint H(u,v,w) \cdot \ (x-u \ \ y-v \ \ z-w)\, du\, dv\, dw$$

where; $V(x,y,z)$ represents the reconstructed three-dimensional volume; $H(u,v,w)$ is the density or intensity at voxel coordinates (u, v, w) in the input image slices; $d(x-u, y-v, z-w)$ is the three-dimensional delta function that localizes the contribution of each voxel in the reconstruction; and the triple integral ($\iiint$) integrates over the entire volume of the input data, combining information from all slices to reconstruct the 3D volume.

This equation plays a crucial role in diagnosis, treatment planning, and surgical guidance in healthcare.

6.7.3. Astronomy and Celestial Navigation

Astronomers leverage 3D geometry to study celestial bodies and their movements. Understanding the spatial relationships between planets, stars, and galaxies allows astronomers to predict astronomical events like eclipses and planetary transits. Furthermore, celestial navigation relies on 3D geometry principles for determining the positions of celestial objects and navigating on Earth (Yair et al., 2003).

In the context of astronomy and celestial navigation, one fundamental equation related to three-dimensional geometry is the equation for the distance between two celestial objects:

Ø **Example: Equation for Distance between Celestial Objects**

$$\text{Distance} = \sqrt{\left((X2-X1)^2+(Y2-Y1)^2+(Z2-Z1)^2\right)}$$

In this equation, $(X1,Y1,Z1)$ represent the three-dimensional coordinates of one celestial object (e.g., a star or planet); $(X2,Y2,Z2)$ represent the three-dimensional coordinates of another celestial object; and distance is the distance between the two celestial objects in three-dimensional space.

This equation is fundamental for various applications, including predicting celestial events, navigating using celestial landmarks, and studying the dynamics of celestial objects in the universe.

6.7.4. Environmental Modeling and Simulation

Environmental scientists use 3D geometry to model natural systems and predict environmental changes. It includes simulating climate patterns, analyzing ocean currents, and studying ecosystems. Researchers gain insights into complex environmental dynamics by representing the Earth's surface and atmospheric conditions in three dimensions (Demitriadou et al., 2020).

Ø **Example: Gaussian Plume Dispersion Equation**

The Gaussian plume dispersion equation is a vital tool in environmental modeling and simulation, used to calculate the concentration of a pollutant at a specific point (x, y, z) in three-dimensional space. It is expressed as follows:

$$C(x,y,z) = \frac{Q}{2\pi u \sigma_y \sigma_z} \cdot \exp\left(-\frac{y^2}{2\sigma_y^2} - \frac{z^2}{2\sigma_z^2}\right)$$

where; $C(x,y,z)$ represents the concentration of the pollutant at the point (x, y, z); (Q) is the emission rate of the pollutant source; (u) is the wind speed; (s_y) and (s_z) are the standard deviations of the pollutant plume in the y and z directions, respectively.

This equation is a fundamental tool for environmental scientists and engineers, allowing them to model the dispersion of pollutants in the atmosphere while considering various factors, including wind conditions, atmospheric stability, and the characteristics of the pollutant source. Its application aids in assessing and mitigating environmental pollution (Oshima et al., 2014).

6.7.5. Aviation and Aerospace Engineering

The aviation and aerospace industries heavily depend on 3D geometry for aircraft design and space exploration. Engineers use geometric principles to calculate aerodynamic properties, design flight trajectories, and ensure the structural integrity of spacecraft and aircraft. Three-dimensional modeling and simulations are crucial for safety and efficiency in these domains.

Ø **Example: Orbital Velocity Equation**

The orbital velocity equation is a fundamental concept in aerospace engineering, used to calculate the velocity required for an object to achieve and maintain a stable orbit around a celestial body. It is expressed as follows:

$$V = \sqrt{\frac{GM}{R}}$$

where; V represents the orbital velocity required for the object; G is the gravitational constant, approximately equal to $\left(6.674\times10^{-11}\right)m^3kg^{-1}s^{-2}$; M is the mass of the celestial body around which the object is orbiting (e.g., Earth or another planet); and R is the radius of the orbit, defined as the distance from the center of the celestial body to the orbiting object.

This equation is critical in mission planning, trajectory calculations, and ensuring the success of space missions and satellite deployments.

In conclusion, three-dimensional geometry transcends theoretical mathematics and finds extensive applications across various disciplines. Its role in engineering, architecture, physics, computer science, medicine, and many other fields is undeniable. This chapter highlights the profound impact of 3D geometry on our modern world and underscores its significance in solving complex real-world problems and advancing technology and science (Oshima et al., 2014).

Activity 6.1

Imagine you're an aerospace engineer designing a new spacecraft. You need to calculate the Euclidean distance between two points in 3D space to determine fuel efficiency. How would you apply the principles of 3D geometry to optimize the spacecraft's trajectory and minimize fuel consumption during its journey to a distant planet?

6.8. SUMMARY

This chapter introduces 3D geometry, making the math behind three-dimensional space simple to grasp. It starts with the basics, explaining the Cartesian coordinate system in 3D and how to find distances and midpoints in space. Then, it dives into equations of lines and planes, showing how they intersect. Vectors in 3D are demystified, revealing their significance in understanding surfaces and solids. The chapter also explores 3D coordinate systems and transformation matrices, making complex transformations understandable. Finally, it shows practical physics, computer science, medicine, astronomy, and engineering applications, making 3D geometry relevant to real-world scenarios.

REFERENCES

1. Alghadari, F., Herman, T., & Prabawanto, S., (2020). Factors affecting senior high school students to solve three-dimensional geometry problems. *International Electronic Journal of Mathematics Education, 15*(3), 0590.
2. Amrouche, C., Bernardi, C., Dauge, M., & Girault, V., (1998). Vector potentials in three-dimensional non-smooth domains. *Mathematical Methods in the Applied Sciences, 21*(9), 823–864.
3. Boileau, P., (1997). The three-dimensional geometry of the proximal humerus: implications for surgical technique and prosthetic design. *The Journal of Bone & Joint Surgery British Volume, 79*(5), 857–865.
4. Cannata, G., Petrelli, C., Barsi, L., & Gallerano, F., (2019). Numerical integration of the contravariant integral form of the Navier–Stokes equations in time-dependent curvilinear coordinate systems for three-dimensional free surface flows. *Continuum Mechanics and Thermodynamics, 31*(2), 491–519.
5. Dabaghi, F. E., & Pironneau, O., (1986). Stream vectors in three-dimensional aerodynamics. *Numerische Mathematik, 48*, 561–589.
6. Darling, W. G., & Gilchrist, L., (1991). Is there a preferred coordinate system for the perception of hand orientation in three-dimensional space? *Experimental Brain Research, 85*, 405–416.
7. Demitriadou, E., Stavroulia, K. E., & Lanitis, A., (2020). Comparative evaluation of virtual and augmented reality for teaching mathematics in primary education. *Education and Information Technologies, 25*, 381–401.
8. Dolgopolov, M. V., & Rodionova, I. N., (2011). Problems involving equations of hyperbolic type in the plane or three-dimensional space with conjugation conditions on a characteristic. *Izvestiya: Mathematics, 75*(4), 681.
9. Faugeras, O., (1995). Stratification of three-dimensional vision: Projective, affine, and metric representations. *JOSA A, 12*(3), 465–484.
10. Fox, R. H., & Artin, E., (1948). Some wild cells and spheres are in three-dimensional space. *Annals of Mathematics, 10*(45), 979–990.
11. Honda, H., Tanemura, M., & Nagai, T., (2004). A three-dimensional vertex dynamics cell model of space-filling polyhedra simulating cell behavior in a cell aggregate. *Journal of Theoretical Biology, 226*(4), 439–453.

12. Huang, J., Li, S., Zi, Y., Qian, Y., Cai, W., Aldén, M., & Li, Z., (2021). Clustering-based particle detection method for digital holography to detect the three-dimensional location and in-plane size of particles. *Measurement Science and Technology*, *32*(5), 055205.
13. Illés, T., Tunyogi-Csapó, M., & Somoskeöy, S., (2011). Breakthrough in three-dimensional scoliosis diagnosis: significance of horizontal plane view and vertebra vectors. *European Spine Journal*, *20*, 135–143.
14. Kondolf, G. M., Boulton, A. J., O'Daniel, S., Poole, G. C., Rahel, F. J., Stanley, E. H., & Nakamura, K., (2006). Process-based ecological river restoration: Visualizing three-dimensional connectivity and dynamic vectors to recover lost linkages. *Ecology and Society*, *11*(2), 9871–9875.
15. Lavaste, F., Skalli, W., Robin, S., Roy-Camille, R., & Mazel, C., (1992). Three-dimensional geometrical and mechanical modelling of the lumbar spine. *Journal of Biomechanics*, *25*(10), 1153–1164.
16. Levit, C., & Lasinski, T., (1991). A tool for visualizing the topology of three-dimensional vector fields. In: *Proc. Visualization'91* (Vol. 1, pp. 33–40).
17. Levitas, V. I., & Preston, D. L., (2002). Three-dimensional Landau theory for multivariant stress-induced martensitic phase transformations. II. Multivariant phase transformations and stress space analysis. *Physical Review B*, *66*(13), 134207.
18. Liu, G. G., Gao, Z., Wang, Q., Xi, X., Hu, Y. H., Wang, M., & Zhang, B., (2022). Topological Chern vectors in three-dimensional photonic crystals. *Nature*, *609*(7929), 925–930.
19. Maxelon, M., & Mancktelow, N. S., (2005). Three-dimensional geometry and tectonostratigraphy of the Pennine zone, Central Alps, Switzerland, and Northern Italy. *Earth-Science Reviews*, *71*(3–4), 171–227.
20. May, K. O., (1966). The impossibility of a division algebra of vectors in three-dimensional space. *The American Mathematical Monthly*, *73*(3), 289–291.
21. Ohtsuki, S., & Tanaka, M., (2006). The flow velocity distribution from the Doppler information on a plane in three-dimensional flow. *Journal of Visualization*, *9*(1), 69–82.
22. Oshima, T., Hiraguri, Y., & Imano, M., (2014). Geometry reconstruction and mesh generation techniques for acoustic simulations over real-life

urban areas using digital geographic information. *Acoustical Science and Technology*, *35*(2), 108–118.

23. Pauwels, P., Van Deursen, D., De Roo, J., Van Ackere, T., De Meyer, R., Van de Walle, R., & Van Campenhout, J., (2011). Three-dimensional information exchange over the semantic web for the domain of architecture, engineering, and construction. *Ai Edam*, *25*(4), 317–332.
24. Picos, K., Diaz-Ramirez, V. H., Kober, V., Montemayor, A. S., & Pantrigo, J. J., (2016). Accurate three-dimensional pose recognition from monocular images using template matched filtering. *Optical Engineering*, *55*(6), 063102–063102.
25. Pogosyan, G. S., Sissakian, A. N., & Vinitsky, S. I., (1994). Interbasis "sphere—cylinder" expansions for the oscillator in the three—dimensional space of constant positive curvature. *Frontiers of Fundamental Physics*, *4*, 429–436.
26. Qiu, Y. C., Yuan, S., Li, X. X., Du, D. Y., Wang, C., Qin, J. S., & Zhou, H. C., (2019). Face-sharing Archimedean solids stacking for the construction of mixed-ligand metal–organic frameworks. *Journal of the American Chemical Society*, *141*(35), 13841–13848.
27. Rathod, H. T., & Rao, H. G., (1995). Integration of polynomials over linear polyhedra in Euclidean three-dimensional space. *Computer Methods in Applied Mechanics and Engineering*, *126*(3–4), 373–392.
28. Sastry, S., Corti, D. S., Debenedetti, P. G., & Stillinger, F. H., (1997). Statistical geometry of particle packings. I. Algorithm for exact determination of connectivity, volume, and surface areas of void space in monodisperse and polydisperse sphere packings. *Physical Review E*, *56*(5), 5524.
29. Sergeev, S. M., Boos, G. E., Mangazeev, V. V., & Stroganov, Y. G., (1996). Ψ-Vectors for three-dimensional models. *Modern Physics Letters A*, *11*(06), 491–498.
30. Silvers, R., (1995). Livemap – a system for viewing multiple transparent and time-varying planes in three dimensional space. In: *Conference Companion on Human Factors in Computing Systems* (Vol. 1, pp. 200–201).
31. Soechting, J. F., & Flanders, M., (1992). Moving in three-dimensional space: frames of reference, vectors, and coordinate systems. *Annual Review of Neuroscience*, *15*(1), 167–191.

32. Xu, D., Lv, L., Ren, J., Shen, W., Wei, S. A., & Dong, L., (2017). Life cycle sustainability assessment of chemical processes: A vector-based three-dimensional algorithm coupled with AHP. *Industrial & Engineering Chemistry Research*, *56*(39), 11216–11227.
33. Yair, Y., Schur, Y., & Mintz, R., (2003). A "thinking journey" to the planets using scientific visualization technologies: Implications to astronomy education. *Journal of Science Education and Technology*, *12*, 43–49.
34. Zhou, J., Liu, M., Wei, C., Gao, C., & Wu, X., (2015). New three-dimensional space vector pulse width modulation of PV-AF system based on the gh $\gamma $ coordinate system. *Turkish Journal of Electrical Engineering and Computer Sciences*, *23*(6), 2017–2029.

CHAPTER 7

Geometrical Models

Contents

7.1. INTRODUCTION

Geometric modeling is a mathematical and computational approach to representing physical objects and their shapes using geometric entities such as points, lines, curves, and surfaces. These geometric entities are defined using mathematical equations and relationships, allowing us to describe and analyze real-world objects precisely and systematically (Bezhanishvili et al., 2022).

Geometric modeling creates a mathematical representation of an object's form and structure. This representation can be two-dimensional (2D) or three-dimensional (3D), depending on the complexity of the object being modeled. Geometric models are the foundation for various applications across multiple fields, where accurately depicting objects and their properties is essential.

In mathematical terms, geometric modeling involves using equations and geometric transformations to describe the coordinates, dimensions, and relationships between points and geometric primitives that compose an object. These equations define the object's geometry, allowing us to visualize, analyze, and manipulate it in a virtual or physical space (Mohammadian & Saidi, 2019). Geometric modeling can be categorized into several types, including wireframe, surface, and solid models, each offering varying levels of complexity and detail in representing physical objects. These models are used extensively in engineering, architecture, computer graphics, and more (Mason, 1997).

Let's explore the significance of geometric modeling in several key fields:

1. **Engineering and Design:** Geometric modeling is central to engineering and design disciplines. Engineers use geometric models to represent mechanical parts, structures, and systems accurately. These models help visualize and test designs before building physical prototypes, saving time and resources. Using geometric equations enables engineers to calculate critical parameters such as dimensions, tolerances, and stress distributions (Jubierre & Borrmann, 2015).

Ø **Example:** A simple example is the equation for calculating the volume of a cylinder (V) with radius (r) and height (h) is given by:

$V = ðr^2h$

2. **Computer Graphics and Animation:** Geometric modeling is the foundation for creating visually compelling digital content in computer graphics and animation. 3D models, defined by geometric equations, represent characters, objects, and environments in video games, movies, and simulations. Using equations for transformations, such as translation and rotation, allows for realistic motion and interactions (Jubierre & Borrmann, 2015).

Ø **Example:** The 3D rotation matrix (R) is used to rotate a point (P) in space based on the angle (è) and axis (u) as follows:

$P' = R(\text{è}, u) \cdot P$

3. **Architecture and Construction:** Architects employ geometric modeling to design buildings, structures, and spaces. Geometric models help architects visualize and communicate their design concepts to clients and builders. These models also facilitate precise measurements and calculations for construction planning (Jubierre & Borrmann, 2015).

Ø **Example:** The Pythagorean theorem $(a^2 + b^2 = c^2)$ is used to calculate the length of the hypotenuse (c) in a right triangle, crucial for determining distances and angles in architectural layouts.

4. **Scientific Research:** Geometric modeling is a critical tool in scientific research across various disciplines. It aids in representing complex natural phenomena, from molecular structures in chemistry to celestial bodies in astronomy. Researchers use equations to model and analyze these systems, gaining insights into their behavior (Yaman et al., 2014).

Ø **Example:** In physics, the equation for the force (F) between two objects with masses m_1 and m_2 separated by a distance (r) is given by the gravitational law:

$$F = G \cdot \frac{m_1 \cdot m_2}{r^2}$$

5. **Manufacturing and Prototyping:** Geometric modeling is indispensable in manufacturing processes. It enables the creation of detailed representations of parts and prototypes using CAD software. These models guide the production of precision components, ensuring quality and compatibility (Desmoulins & Van Labeke, 1996).

Ø **Example:** Geometric dimensioning and tolerancing (GD&T) uses equations to define the allowable variations in size and geometry of manufactured parts, ensuring they meet specified requirements.

7.2. GEOMETRIC ABSTRACTION

Abstraction in geometry means taking complex things and making them easier to work with by focusing on essential details (De Jong, 1996).

Did You Know?

Geometric abstraction in modeling can be traced back to ancient civilizations like the Egyptians and Greeks, who used geometric principles in their architectural and artistic creations. This enduring tradition highlights the timeless utility of geometric abstraction in diverse fields.

Imagine you have a round table. In geometric abstraction, you might represent it as a perfect circle. This simplification helps us use math to understand and work with shapes more effectively.

Ø **Example:** The equation for a circle with a center (C) and radius (r) is given by:

$$(x - C_x)^2 + (y - C_y)^2 = r^2$$

7.2.1. Role of Geometric Abstraction in Modeling

Geometric abstraction is pivotal in various fields because it can simplify complex real-world objects while retaining essential geometric properties. Its significance is evident in several practical applications (Langer, 1937):

1. **Simplifying Complex Shapes:** Consider modeling a car's tire, which can be abstracted as a cylinder. Such simplification allows the use of equations like the surface area formula $(A = 2pr^2 + 2prh)$

to precisely describe and analyze it. This equation defines the surface area (A) of a cylinder with a radius (r) and height (h).

2. **Precision and Efficiency:** Geometric abstraction enables precise mathematical descriptions, ensuring accurate measurements and calculations. This precision is particularly valuable in engineering and design disciplines, where exactness is critical.
3. **Visualization:** Geometric models are powerful tools for visualizing complex objects and designs. Architects and engineers employ them to gain insights into structural intricacies, often visualizing building components as geometric shapes.
4. **Analysis:** The mathematical tractability of simplified geometric models empowers engineers to conduct in-depth analyses of properties such as volume, area, and strength. These analyses are vital for ensuring structural integrity and functionality.
5. **Communication:** Geometric abstraction provides a universal language that transcends professional boundaries. It facilitates the clear exchange of design plans and ideas among architects, engineers, and designers (Langer, 1937).

7.2.2. Real-World vs. Geometric Simplification

Let's explore the distinction between real-world complexity and geometric simplification by considering specific examples with equations (Mueller, 2003):

1. **Real-World Complexity:** Real-world objects such as trees with countless leaves exhibit intricate details and irregularities. Geometric abstraction simplifies such complexities by representing objects like trees as simplified geometric shapes like cones or cylinders. These shapes are succinctly defined by equations, facilitating precise description and analysis. For instance, the volume of a cone with base radius (r) and height (h) is calculated using the equation $\left(V = \frac{1}{3}\eth r^2 h\right)$.
2. **Real-World Details:** Natural landscapes, such as mountain ranges, feature rugged terrain with varying heights and shapes, representing a level of detail that can be overwhelming. In geometric modeling, mountains are often simplified using

triangular shapes. This simplification makes calculations like elevation and slope more manageable.

3. **Real-World Fluids:** Visualize the intricate dynamics of flowing water with its ripples and currents. Geometric models may represent a river's path as a smooth curve, simplifying the dynamic nature of fluid flow while maintaining practical accuracy. For instance, modeling structural elements in architectural design often simplifies intricate designs into geometric shapes like rectangles and circles. These shapes are characterized by equations that allow for precise calculations of dimensions and proportions (Dunlop, 2009).

In conclusion, geometric abstraction is a powerful tool in mathematical modeling, striking a harmonious balance between simplification and precision. It empowers professionals across various disciplines to efficiently represent, analyze, and communicate complex real-world objects, bridging the gap between intricate reality and mathematical representation.

7.3. TYPES OF GEOMETRIC MODELS

The types of the geometric models are given in subsections (Withers, 1905).

7.3.1. Wireframe Models

Wireframe models are a fundamental type of geometric representation used in computer graphics and engineering. They serve as a simplified framework for visualizing three-dimensional (3D) objects. In a wireframe model, the object's surface is represented by lines and curves, creating a skeletal structure that outlines its shape. These models depict an object's geometry, facilitating visualization, analysis, and design in various fields (Hempel, 1945).

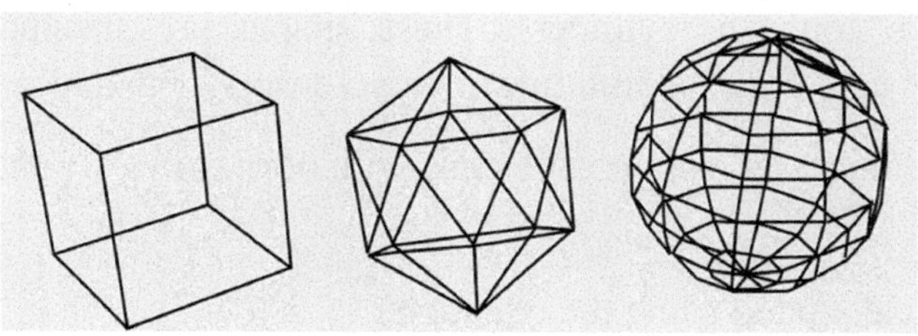

Figure 7.1. Illustration of shapes in wireframe model.

Source: Anyrgb, Creative Commons License.

Wireframe models use geometric equations to define the relationships between points, lines, and curves. One fundamental equation used in wireframe modeling is the equation of a line in 3D space:

$$r(t) = r_0 + tv$$

here; $r(t)$ represents a point on the line (r_0) is the line's origin; (v) is the direction vector; and (t) is a parameter that varies along the line. The equation allows us to define and trace lines in 3D space.

7.3.1.1. Example of Wireframe Models

Consider a simple wireframe model of a cube. In this case, each cube edge is represented by a line segment. The vertices (corners) of the cube are defined by their 3D coordinates, which can be expressed as points in space. Let's take one edge of the cube, connecting two vertices, and represent it using a wireframe equation (Gray, 2015):

- **Vertex A:** $A(x_1, y_1, z_1)$
- **Vertex B:** $B(x_2, y_2, z_2)$

The equation of the line segment connecting A and B is given by:

$$r(t) = A + t(B - A)$$

Using this equation, we can calculate and plot all the points along this edge of the cube. We construct a wireframe representation of the entire cube by repeating this process for all edges.

7.3.1.2. Applications of Wireframe Models

Wireframe models find applications in various fields, including architecture, engineering, and computer graphics. Architects use wireframes to visualize building structures, engineers employ them to design machinery and mechanical parts, and computer graphics artists use them as the foundation for creating 3D models in video games and animations. These models offer a simplified yet effective way to convey geometric information (Eves et al., 1969).

7.3.2. Surface Models

Unlike wireframe models that primarily represent the edges of objects, surface models provide a more detailed representation by defining the object's outer shell or surface. These models allow for a more accurate portrayal of object geometry, enabling realistic visualizations, simulations, and analyses.

Remember

Surface models are particularly valuable in computer graphics and gaming industries, where they are used to create realistic 3D objects. Think of surface models as the skin that covers wireframe structures, providing depth and texture to digital objects.

Surface models rely on equations to define the mathematical functions that describe the shape of the object's surface. These equations use geometric terms to represent points on the surface. One common equation for surface modeling is the parametric equation of a surface in 3D space (Michael, 1988):

$$r(u,v) = x(u,v), y(u,v), z(u,v)$$

here; $r(u,v)$ represents a point on the surface; and (u) and (v) are parameters that vary over a specified domain. The functions $x(u,v)$, $y(u,v)$, and $z(u,v)$ define the position of the point in 3D space as a function of the parameters (u) and (v). These equations allow for the precise definition of complex surfaces.

7.3.2.1. Example of Surface Models

Consider a simple example of a surface model: a parametric representation of a sphere. The parametric equations for a sphere with radius (r) are as follows:

$$x(u,v) = r\sin(u)\cos(v)$$

$$y(u,v) = r\sin(u)\sin(v)$$

$$z(u,v) = r\cos(u)$$

In these equations, (u) and (v) represent angles that vary within specific ranges. As (u) and (v) change, the equations calculate the coordinates (x, y, z) of points on the sphere's surface. By varying (u) and (v) over their respective domains, we can generate a complete representation of the sphere's surface (Heis et al., 2020).

7.3.2.2. Applications of Surface Models

Surface models are extensively used in computer graphics, engineering, and scientific simulations. In computer graphics, 3D models of objects, characters, and environments are often represented using surface modeling techniques. Engineers use surface models to design complex mechanical parts and visualize their geometry. Scientists employ them to simulate physical phenomena and study surfaces, such as terrain modeling for environmental analysis (Singh, 2021).

7.3.3. Solid Models

Solid models are a fundamental geometric representation employed extensively in engineering, design, and manufacturing. Unlike surface models that focus on the outer shell of objects, solid models offer a complete representation of 3D objects, including their interior volume. These models are crucial for precise and comprehensive modeling of physical objects, enabling detailed analysis, prototyping, and simulations (Ungar, 2008).

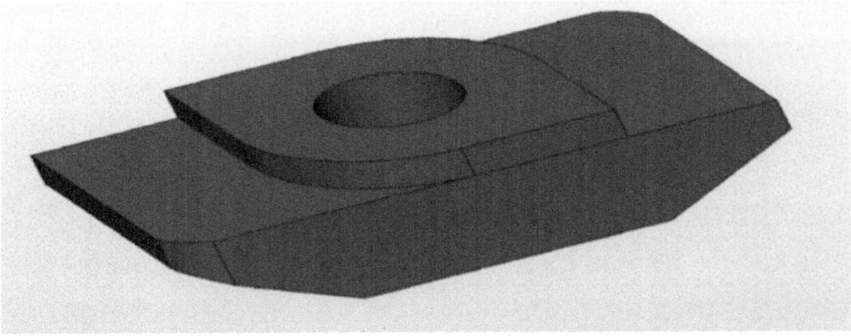

Figure 7.2. Example of a solid model geometry.

Source: Prescient Technologies, Creative Commons License.

Solid models rely on equations that define the geometry of the object's surface and interior. These equations incorporate geometric terms and functions to describe the shape and properties of the solid. One common equation used in solid modeling is the mathematical representation of a sphere:

$$x^2 + y^2 + z^2 = r^2$$

here; $(x),(y)$, and (z) are coordinates in 3D space; and (r) represents the radius of the sphere. This equation defines all the points. (x, y, z) that belong to the solid sphere on its surface and within its interior volume (Boutry et al., 2015).

7.3.3.1. Example of Solid Models

Consider a practical example of a solid model: a cube representation. Unlike surface models representing only the cube's outer shell, a solid model includes its exterior faces and interior volume. The equations for a cube's solid model are:

$$-a/2 \leq x \leq a/2$$

$$-b/2 \leq y \leq b/2$$

$$-c/2 \leq z \leq c/2$$

In these equations, $(a),(b)$, and (c) represent the lengths of the cube's sides along the x, y, and z axes, respectively. These equations define the cube's solid's boundaries, encompassing its surface and interior.

7.3.3.2. Applications of Solid Models

Solid models find wide-ranging applications in engineering, manufacturing, architecture, and scientific research. Engineers use solid models to design complex machinery, analyze stress distribution within structures, and perform simulations for various physical systems. Architects employ them to create precise building plans, and manufacturers rely on solid models for prototyping and production. Additionally, solid models are essential in medical imaging for representing anatomical structures in 3D (Christiansen, 2013).

7.3.4. Other Specialized Models

Some specialized models are discussed in subsections.

7.3.4.1. Point Cloud Models

Point cloud models represent 3D objects as a collection of discrete points in space, where each point is defined by its 3D coordinates. These models are commonly used in 3D scanning and digital representation of physical objects. The equation for a point cloud model is straightforward:

$$P(x, y, z)$$

here; ( ) represents a point with coordinates. (x, y, z). Point cloud models do not explicitly define surfaces or volumes but provide a cloud of points collectively representing the object (Freudenthal, 1966).

- Ø **Example of Point Cloud Models:** Consider the scanning of a statue using a 3D laser scanner. The scanner records the position of numerous points on the statue's surface, creating a point cloud model. Each point (x, y, z) in the model corresponds to a specific location on the statue. The point cloud accurately captures the statue's shape, which can later be reconstructed or analyzed.

7.3.4.2. NURBS Models

Non-Uniform Rational B-Spline (NURBS) models are mathematical representations in CAD and 3D modeling. NURBS models define surfaces and curves using B-spline functions, expressed mathematically. The equations for NURBS models involve control points, weights, and basic functions. A NURBS curve equation, for example, is represented as (Moussard, 2012):

$$C(u) = \sum_{i=0}^{n} N_i(u) P_i$$

here; $C(u)$ is the curve; (u) is the parameter; (n) is the degree of the curve; $N_i(u)$ are the basis functions; and (P_i) are the control points. NURBS models offer precise control over the shape and smoothness of curves and surfaces.

- Ø **Example of NURBS Models:** In automotive design, NURBS models create car bodies' smooth, flowing surfaces. The NURBS equation's control points and basic functions allow designers to precisely define the shape of curves, ensuring aerodynamic and aesthetic qualities.

7.3.4.3. Parametric Models

Parametric models use mathematical equations with parameters to define geometric shapes and structures. These flexible models allow designers to modify objects by adjusting the parameters while maintaining mathematical consistency. An example of a parametric equation is the equation of a parametric circle (Stojanović et al., 2011):

$$x(t) = r\cos(t)$$

$$y(t) = r\sin(t)$$

here; $x(t)$ and $y(t)$ represent the coordinates of points on the circle; (r) is the radius; and (t) is a parameter that varies to trace out the circle.

Ø **Example of Parametric Models:** Architects often use parametric models to design buildings

with complex and variable shapes. By adjusting parameters in equations that define building components, architects can explore various design possibilities while ensuring structural integrity.

7.3.4.4. Fractal Models

Fractal models are mathematical constructs that exhibit self-similarity at different scales. They are commonly used in natural phenomena simulations and artistic representations. The most famous fractal, the Mandelbrot Set, is defined by an iterative equation (Russell, 1903):

$$Z_{n+1} = Z_n^2 + C$$

here; (Z_n) is a complex number; and (C) is a constant. The Mandelbrot Set is generated by iterating This equation, and the resulting points are plotted to create intricate and self-similar patterns.

Ø **Example of Fractal Models:** Fractal models find applications in generating realistic terrain for computer games. By applying fractal algorithms, designers can create terrain that exhibits the same geological features at various levels of detail, from mountains to tiny rock formations.

In summary, specialized geometric models, including point clouds, NURBS, parametric, and fractal models, offer unique mathematical approaches to represent and describe complex shapes and structures. These models find applications in various fields, from 3D scanning to architectural design, and are valuable tools for precise representation and analysis (Puitg & Dufourd, 2000).

7.4. THE ROLE OF COMPUTERS IN GEOMETRIC MODELING

The role of computers in geometric modeling is pivotal, transforming how we design, analyze, and visualize complex geometries. Using mathematical equations, computers enable precise and efficient representation of intricate shapes, both in 2D and 3D.

Whether in architecture, engineering, or scientific simulations, computers facilitate the creation of digital models that offer unparalleled precision and flexibility. They empower designers and engineers to iterate rapidly, simulate physical phenomena, and visualize intricate structures, revolutionizing industries and enhancing our ability to bring innovative ideas to life (Freudenthal, 1971).

Tip

Geometric abstraction in modeling can be traced back to ancient civilizations like the Egyptians and Greeks, who used geometric principles in their architectural and artistic creations. This enduring tradition highlights the timeless utility of geometric abstraction in diverse fields.

7.4.1. Evolution of Computer-Aided Design (CAD)

Computer-aided design (CAD) is the backbone of modern geometric modeling, revolutionizing how we design, visualize, and analyze complex geometries. CAD software has a rich history, evolving to become a critical tool in various industries (Winer & Battista, 2022).

The fundamental equation underlying CAD is the mathematical representation of geometric shapes. Shapes can be represented using parametric equations that define their form and properties. For instance, a parametric equation for a circle is:

$$x(t) = r\cos(t)$$

$$y(t) = r\sin(t)$$

here; $x(t)$ and $y(t)$ describe the circle's coordinates; (r) is the radius; and (t) is a parameter that traces the circle. CAD software employs similar equations to create and manipulate geometric entities (Boutry et al., 2019).

The evolution of CAD can be summarized in three stages:

1. **Manual Drafting:** Before computers, designs were created manually, often involving tedious hand-drawn sketches and blueprints. Geometrical tools like compasses and rulers were essential.
2. **2D CAD:** The advent of computers introduced 2D CAD software, enabling engineers and designers to create precise 2D drawings. Geometric terms like lines, arcs, and polygons were employed in these digital representations.
3. **3D CAD:** The most significant advancement was the transition to 3D CAD, allowing the creation of three-dimensional models of objects and structures. This stage expanded the use of equations to define complex 3D surfaces and solids.

7.4.2. Digital Representation of Geometry

Digital representation of geometry involves encoding geometric shapes and structures into a computer-readable format. This representation relies heavily on mathematical equations and geometrical terms to accurately describe the geometry. Some essential aspects of digital representation include (Toussaint, 2015):

1. **Parametric Equations:** As discussed earlier, parametric equations play a central role in digital representation. They define the relationships between points, curves, and surfaces, allowing for precise and dynamic modeling.
2. **Vector Graphics:** Vector graphics use mathematical equations to describe shapes through connected points, lines, and curves. These equations enable scalability without loss of detail.
3. **Polygonal Meshes:** Complex surfaces are often represented using polygonal meshes. Each polygon, typically a triangle, is defined by vertices with specific coordinates. Equations for triangles are employed to calculate properties like surface area and normal vectors.

4. **Boundary Representation (B-Rep):** In 3D modeling, B-Rep defines objects by their boundaries, such as faces, edges, and vertices. These boundaries are mathematically defined, often using equations.

7.5. PRACTICE PROBLEMS

Some practical problems related to the geometrical modeling are discussed in subsections (Chen & Wang, 2013).

7.5.1. Parametric Equations

Given the parametric equations for a circle:

$$x(t) = R\cos(t)$$

$$y(t) = R\sin(t)$$

- Find the coordinates of points on the circle when $(t ð/4\ .)$
- Calculate the radius (R) when $(x=3)$ and $(y=4)$.

Ø **Solution 1:**

- When $(t ð/4\ ,)$ we have:

$$x(ð/4) = R\cos(ð/4) = R\cdot\sqrt{2}/2$$

$$y(ð/4) = R\sin(ð/4) = R\cdot\sqrt{2}/2$$

So, the coordinates of the point on the circle at $(t = ð/4)$ *are* $\left(R\cdot\sqrt{2}/2, R\cdot\sqrt{2}/2\right)$.

- To find the radius (R), we use the Pythagorean theorem:

$$x^2 + y^2 = R^2$$

$$3^2 + 4^2 = R^2$$

$$9 + 16 = R^2$$

$$R^2 = 25$$

$$R = 5$$

7.5.2. Matrix Transformation

Apply a 2D transformation matrix to the point $(2,3)$ with the following matrix:

$$T = \begin{bmatrix} 0.866 & -0.5 \\ 0.5 & 0.866 \end{bmatrix}$$

- Find the new coordinates of the point after transformation.
- Determine the type of transformation represented by a matrix (T).

Ø **Solution 2:**

- To find the new coordinates, multiply the point by the transformation matrix:

$$\begin{bmatrix} 0.866 & -0.5 \\ 0.5 & 0.866 \end{bmatrix} \begin{bmatrix} 2 \\ 3 \end{bmatrix} = \begin{bmatrix} 0.866 \cdot 2 - 0.5 \cdot 3 \\ (0.5 \cdot 2 + 0.866 \cdot 3) \end{bmatrix} = \begin{bmatrix} 0.732 \\ 3.232 \end{bmatrix}$$

So, the new coordinates after transformation are $(0.732, 3.232)$.

- Matrix (T) represents a rotation transformation by an angle of $(30)\mathit{degrees}\left(\mathit{or}\,(\pi/6)\mathit{radians}\right)$ counterclockwise.

7.5.3. Circle Equation

Find the equation of a circle with its center at $(1,-2)$ and a radius of (4).

Ø **Solution 3:** The equation of a circle with a center (h,k) and radius (r) is given by:

$$(x-h)^2 + (y-k)^2 = r^2$$

Plugging in the given values:

$$(x-1)^2 + (y+2)^2 = 4^2$$

$$(x-1)^2 + (y+2)^2 = 16$$

So, the equation of the circle is $(x-1)^2 + (y+2)^2 = 16$.

7.5.4. Line Equation

Find the equation of the line passing through points $(2,3)$ and $(4,5)$.

Ø **Solution 4:** The equation of a line passing through points (x_1, y_1) and (x_2, y_2) is given by the point-slope form:

$$y - y_1 = \frac{y_2 - y_1}{x_2 - x_1}(x - x_1)$$

Plugging in the given points:

$$y - 3 = \frac{5-3}{4-2}(x-2)$$

$$y - 3 = \frac{2}{2}(x-2)$$

$$y - 3 = x - 2$$

So, the equation of the line is $(y = x + 1)$.

7.5.5. Area of a Triangle

Calculate the area of a triangle with vertices at $(1,2), (4,6)$, and $(7,2)$.

Ø **Solution 5:** The area (A) of a triangle with vertices $(x_1, y_1), (x_2, y_2)$, and (x_3, y_3) can be calculated using the shoelace formula:

$$A = \frac{1}{2}\left|x_1(y_2 - y_3) + x_2(y_3 - y_1) + x_3(y_1 - y_2)\right|$$

Plugging in the given coordinates:

$$A = \frac{1}{2}\left|1(6-2) + 4(2-2) + 7(2-6)\right|$$

$$A = \frac{1}{2}|4-8-20|$$

$$A = \frac{1}{2}|-24|$$
$$A=12$$

So, the area of the triangle is (12).

7.5.6. Intersection of Lines

Determine if the lines $(2x-3y=7)$ and $(4x+5y=9)$ intersect, and if so, find the point of intersection.

Ø **Solution 6:** To check if two lines intersect, we can solve the system of equations formed by the lines:

$$2x-3y=7$$

$$4x+5y=9$$

Solving this system yields:

$$x=2$$

$$y=-1$$

So, the lines intersect at the point $(2,-1)$.

7.5.7. Area of a Circle Sector

Calculate the area of a sector of a circle with a central angle of (60°) and a radius of (8).

Ø **Solution 7:** The area (A_{sector}) of a sector of a circle with central angle (è) and radius (r) can be calculated using the formula:

$$A_{\text{sector}} = \frac{è}{360^\circ} ðr^2$$

Plugging in the given values:

$$A_{\text{sector}} = \frac{60^\circ}{360^\circ} ð \cdot 8^2$$

$$A_{\text{sector}} = \frac{1}{6} ð \cdot 64$$

$$A_{\text{sector}} = \frac{32}{3} ð$$

So, the area of the sector is $\left(\frac{32}{3} ð\right)$.

7.5.8. Volume of a Cone

Calculate the volume of a cone with a height of (10) units and a base radius of (4) units.

Ø **Solution 8:** The volume (V) of a cone with height (h) and base radius (r) can be calculated using the formula:

$$V = \frac{1}{3} ð r^2 h$$

Plugging in the given values:

$$V = \frac{1}{3} ð \cdot 4^2 \cdot 10$$

$$V = \frac{1}{3} ð \cdot 160$$

$$V = \frac{160}{3} ð$$

So, the volume of the cone is $\left(\frac{160}{3} ð\right)$.

7.5.9. Equation of a Parabola

Find the equation of a parabola with its vertex at $(3,-2)$ and its focus at $(3,1)$.

Ø **Solution 9:** The equation of a parabola with vertex (h,k) and focus $(h,k+p)$ is given by:

$$(x-h)^2 = 4p(y-k)$$

In this case, $(h=3)$ and $(k=-2)$. The distance from the vertex to the focus is $p=1-(-2)=3$.

Plugging in these values:

$$(x-3)^2 = 4\cdot 3(y-(-2))$$

$$(x-3)^2 = 12(y+2)$$

So, the equation of the parabola is $(x-3)^2 = 12(y+2)$.

7.5.10. Intersection of Planes

Determine if the planes $(2x+3y-z=4)$ and $(x-2y+3z=5)$ intersect, and if so, find the line of intersection.

Ø **Solution 10:** To check if two planes intersect, we can solve the system of equations formed by the planes:

$$2x+3y-z=4$$

$$x-2y+3z=5$$

Solving this system yields:

$$x=1$$

$$y=2$$

$$z=3$$

So, the planes intersect at the point $(1,2,3)$, and the line of intersection is defined by these equations.

Activity 7.1

Imagine you're an architect tasked with designing a futuristic building with intricate geometric patterns. Explain how the use of wireframe, surface, and solid models in computer-aided design (CAD) can assist you in visualizing and perfecting your design before construction begins. Additionally, discuss the role of parametric equations in fine-tuning complex geometric details.

7.6. SUMMARY

This chapter introduces the world of geometric modeling, where simplifying complex real-world objects into geometric shapes is crucial. It discusses various geometric models, including wireframe, surface, and solid models, illustrating their practical applications. Furthermore, it highlights the role of computers in geometric modeling and the evolution of CAD. The chapter also offers practical problems, emphasizing the significance of understanding parametric equations and matrix transformations in creating precise digital geometry representations. Geometric abstraction, rooted in ancient civilizations, remains a timeless tool in diverse fields, showcased in this chapter's exploration of geometric models.

REFERENCES

1. Acaroglu, O., Ozdemir, L., & Asbury, B., (2008). A fuzzy logic model to predict specific energy requirement for TBM performance prediction. *Tunnelling and Underground Space Technology*, *23*(5), 600–608.
2. Adamatzky, A., Kitson, S., Costello, B. D. L., Matranga, M. A., & Younger, D., (2011). Computing with liquid crystal fingers: Models of geometric and logical computation. *Physical Review E*, *84*(6), 061702.
3. Amoroso, R. L., & Rauscher, E. A., (2009). *Holographic Anthropic Multiverse, The: Formalizing the Complex Geometry of Reality* (Vol. 43). World Scientific.
4. Bezem, M., & Coquand, T., (2004). Newman's Lemma – a case study in proof automation and geometric logic. In: *Current Trends in Theoretical Computer Science: The Challenge of the New Century – Vol. 1: Algorithms and Complexity – Vol. 2: Formal Models and Semantics* (pp. 267–282).
5. Bezhanishvili, N., Ciancia, V., Gabelaia, D., Grilletti, G., Latella, D., & Massink, M., (2022). Geometric model checking of continuous space. *Logical Methods in Computer Science, 18, 1-38.*
6. Boutry, P., Braun, G., & Narboux, J., (2019). Formalization of the arithmetization of Euclidean plane geometry and applications. *Journal of Symbolic Computation*, *90*, 149–168.
7. Boutry, P., Narboux, J., & Schreck, P., (2015). *Parallel Postulates and Decidability of Intersection of Lines: A Mechanized Study Within Tarski's System of Geometry, 1, 1-58.*
8. Chen, X., & Wang, D., (2013). Formalization and specification of geometric knowledge objects. *Mathematics in Computer Science*, *7*, 439–454.
9. Christiansen, A., (2013). The understanding of parallel lines in early 19th century textbooks: A comparison between two Norwegian geometry books from 1,827 and 1,835. *Dig Where You Stand, 3, 125-145.*
10. De Jong, W. R., (1996). *Gottlob Frege and the Analytic-Synthetic Distinction Within the Framework of the Aristotelian Model of Science, 1, 1-15.*
11. Desmoulins, C., & Van Labeke, N., (1996). Towards student modelling in geometry with inductive logic programming. In: *Proceedings of the European Conference on Artificial Intelligence in Education* (Vol. 1, pp. 105–111).

12. Dunlop, K., (2009). Why Euclid's geometry brooked no doubt: JH Lambert on certainty and the existence of models. *Synthase, 167*, 33–65.
13. Đurđević, S. S., Narboux, J., & Janičić, P., (2015). Automated generation of machine verifiable and readable proofs: A case study of Tarski's geometry. *Annals of Mathematics and Artificial Intelligence, 74*(3–4), 249–269.
14. Eves, H., Rolwing, R. H., & Levine, M., (1969). Historically speaking – the parallel postulate. *The Mathematics Teacher, 62*(8), 665–669.
15. Fernández, F. A., Camacho, F. G., Pérez, J. S., Sevilla, J. F., & Grima, E. M., (1997). A model for light distribution and average solar irradiance inside outdoor tubular photobioreactors for the microalgal mass culture. *Biotechnology and Bioengineering, 55*(5), 701–714.
16. Frege, G., (1980). *The Foundations of Arithmetic: A Logico-Mathematical Enquiry into the Concept of Number*. Northwestern University Press.
17. Freudenthal, H., (1966). The main trends in the foundations of geometry in the 19th century. In: *Studies in Logic and the Foundations of Mathematics* (Vol. 44, pp. 613–621). Elsevier.
18. Freudenthal, H., (1971). Geometry between the devil and the deep sea. In: *The Teaching of Geometry at the Pre-College Level: Proceedings of the Second CSMP International Conference Co-Sponsored by Southern Illinois University and Central Midwestern Regional Educational Laboratory* (pp. 137–159). Dordrecht: Springer Netherlands.
19. Gan, W., Yu, X., & Wang, M., (2019). Automatic understanding and formalization of plane geometry proving problems in natural language: A supervised approach. *International Journal on Artificial Intelligence Tools, 28*(04), 1940003.
20. Gladstone, R. M., Lee, V., Rougier, J., Payne, A. J., Hellmer, H., Le Brocq, A., & Cornford, S. L., (2012). Calibrated prediction of Pine Island Glacier retreat during the 21st and 22nd centuries with a coupled flowline model. *Earth and Planetary Science Letters, 333*, 191–199.
21. Gray, J., (2015). A note on lines and planes in Euclid's geometry. In: *Mathematizing Space: The Objects of Geometry from Antiquity to the Early Modern Age* (pp. 65–73). Springer International Publishing.
22. Grosholz, E., (2000). Frege and the surprising history of logic: Introduction to Claude Imbert, "Gottlob Frege, one more time." *Hypatia, 15*(4), 151–155.

23. Halsted, G. B., (1900). Gauss and the non-Euclidean Geometry. Carl Friedrich Gauss Werke. Band VIII. Göttingen. 1900, p. 458. *Science, 12*(309), 842–846.
24. Hannesdóttir, H., Björnsson, H., Pálsson, F., Aðalgeirsdóttir, G., & Guðmundsson, S., (2015). Variations of southeast Vatnajökull ice cap (Iceland) 1650–1900 and reconstruction of the glacier surface geometry at the Little Ice Age maximum. *Geografiska Annaler: Series A, Physical Geography*, *97*(2), 237–264.
25. Hansen, M., (1993). What is geometry? *Mathematics in School*, *22*(4), 9–11.
26. Heis, J., Rechter, O., & Posy, C., (2020). Kant on parallel lines: Definitions, postulates, and axioms. *Kant's Philosophy of Mathematics*, *1*, 157–80.
27. Hempel, C. G., (1945). Geometry and empirical science. *The American Mathematical Monthly*, *52*(1), 7–17.
28. Herbst, K., Kopp, A., & Heber, B., (2013). Influence of the terrestrial magnetic field geometry on the cutoff rigidity of cosmic ray particles. In: *Annales Geophysicae* (Vol. 31, No. 10, pp. 1637–1643). Göttingen, Germany: Copernicus Publications.
29. Iseri, H., (2002). Partially paradosist Smarandache geometries. *Smarandache Notions Journal*, *13*, 5–13.
30. Islam, M. R., & Ali, M., (2010). Elevation plane beam scanning of a novel parasitic array radiator antenna for 1,900 MHz mobile handheld terminals. *IEEE Transactions on Antennas and Propagation*, *58*(10), 3344–3352.
31. Jackson, H., (1981). Eike-Henner W. Kluge – Introduction – On the foundations of geometry and formal theories of arithmetic, by Gottlob Frege, translated, and with an introduction by Eike-Henner W. Kluge, Yale University Press, New Haven and London (1971, pp. xi–xlii). -Gottlob Frege – Letter from G. Frege to Heinrich Liebmann – On the foundations of geometry and formal theories of arithmetic, by Gottlob Frege, translated, and with an introduction by Eike-Henner W. Kluge, Yale University Press, New Haven and London (1971, pp. 3–5). *The Journal of Symbolic Logic*, *46*(1), 175–179.
32. Jubierre, J. R., & Borrmann, A., (2015). Knowledge-based engineering for infrastructure facilities: assisted design of railway tunnels based on logic models and advanced procedural geometry dependencies.

Journal of Information Technology in Construction (ITcon), *20*(26), 421–441.

33. Kim, Y. H., Choe, K. U., & Ri, R. K., (2019). Application of fuzzy logic and geometric average: A Cu sulfide deposits potential mapping case study from Kapsan Basin, DPR Korea. *Ore Geology Reviews*, *107*, 239–247.
34. Kimura, F., Suzuki, H., Ando, H., Sato, T., & Kinosada, A., (1987). Variational geometry based on logical constraints and its applications to product modelling. *CIRP Annals*, *36*(1), 65–68.
35. Klein, R., (1998). The role of constraints in geometric modelling. In: *Geometric Constraint Solving and Applications* (pp. 3–23). Berlin, Heidelberg: Springer Berlin Heidelberg.
36. Langer, S. K., (1937). Heinrich Scholz and Friedrich Bachmann – Der wissenschaftliche Nachlass von Gottlob Frege. Actes du Congrès International de Philosophie Scientifique, VIII Histoire de la logique et de la philosophie scientiftque, Actualités scientifiques et industrielles 395, Hermann & Cie, Paris (1936, pp. 24–30). *The Journal of Symbolic Logic*, *2*(1), 56, 57.
37. Lee, H. D. P., (1935). Geometrical method and Aristotle's account of first Principles. *The Classical Quarterly*, *29*(2), 113–124.
38. Ma, S., Shi, Z., Shao, Z., Guan, Y., Li, L., & Li, Y., (2016). Higher-order logic formalization of conformal geometric algebra and its application in verifying a robotic manipulation algorithm. *Advances in Applied Clifford Algebras*, *26*, 1305–1330.
39. Magaud, N., Narboux, J., & Schreck, P., (2012). A case study in formalizing projective geometry in Coq: Desargues theorem. *Computational Geometry*, *45*(8), 406–424.
40. Malt, P., & Thierry, S. E., (2010). A formalization of geometric constraint systems and their decomposition. *Formal Aspects of Computing*, *22*, 129–151.
41. Mason, M. M., (1997). The Vanx Hiele model of geometric understanding and mathematically talented students. *Journal for the Education of the Gifted*, *21*(1), 38–53.
42. McClymont, A. F., Hayashi, M., Bentley, L. R., & Christensen, B. S., (2013). Geophysical imaging and thermal modeling of subsurface morphology and thaw evolution of discontinuous permafrost. *Journal of Geophysical Research: Earth Surface*, *118*(3), 1826–1837.

43. Meikle, L. I., & Fleuriot, J. D., (2003). Formalizing Hilbert's Grundlagen in Isabelle/Isar. In: *International Conference on Theorem Proving in Higher Order Logics* (pp. 319–334). Berlin, Heidelberg: Springer Berlin Heidelberg.
44. Michael, J. S., (1988). Beltrami's model and the independence of the parallel postulate. *History and Philosophy of Logic*, *9*(1), 13–34.
45. Mlodinow, L., (2010). *Euclid's Window: The Story of Geometry from Parallel Lines to Hyperspace*. Simon and Schuster, 1, 1-19.
46. Mohammadian, P. K., & Saidi, M. H., (2019). Simulation of startup operation of an industrial twin-shaft gas turbine based on geometry and control logic. *Energy*, *183*, 1295–1313.
47. Mölg, T., Hardy, D. R., & Kaser, G., (2003). Solar-radiation-maintained glacier recession on Kilimanjaro drawn from combined ice-radiation geometry modeling. *Journal of Geophysical Research: Atmospheres*, *108*(D23).
48. Moussard, G., (2012). The notion of method in 19th century French geometry teaching: Three textbooks. In: *Dig Where You Stand 2: Proceedings of the Second International Conference on the History of Mathematics Education* (pp. 333–350).
49. Mueller, I., (2003). Remarks on Euclid's Elements I, 32 and the parallel postulate. *Science in Context*, *16*(3), 287–297.
50. Mumma, J., (2008). Nathaniel Miller – Euclid and his 20th century rivals: Diagrams in the logic of Euclidean geometry. *CSLI Studies in the Theory and Applications of Diagrams 16(2), 256-264.*
51. Pomarlan, M., & Bateman, J. A., (2020). Embodied functional relations: A formal account combining abstract logical theory with grounding in simulation. In: *FOIS* (pp. 155–168).
52. Potaga, A., (2002). The logical objectivism' of Gottlob Frege: A possible answer as to the applicability of mathematics. *Philosophical Inquiry*, *24*(3–4), 77–94.
53. Puitg, F., & Dufourd, J. F., (2000). Formalizing mathematics in higher-order logic: A case study in geometric modelling. *Theoretical Computer Science*, *234*(1–2), 1–57.
54. Reck, E. H., (1992). *FREGE, Wittgenstein, and Platonism in Mathematics (Gottlob Frege, Ludwig Wittgenstein)*. Doctoral dissertation, The University of Chicago, 1, 1-16.

55. Remo, J. W., Pinter, N., & Heine, R., (2009). The use of retro-and scenario-modeling to assess effects of 100+ years river of engineering and land-cover change on Middle and Lower Mississippi River flood stages. *Journal of Hydrology*, *376*(3–4), 403–416.
56. Richardson, A., (2003). The geometry of knowledge: Lewis, Becker, Carnap, and the formalization of philosophy in the 1920s. *Studies in History and Philosophy of Science Part A*, *34*(1), 165–182.
57. Roanes-Lozano, E., Laita, L. M., & Roanes-Macías, E., (1998). A polynomial model for multi-valued Logics with a touch of algebraic geometry and computer algebra. *Mathematics and Computers in Simulation*, *45*(1–2), 83–99.
58. Rowe, D. E., (2013). Mathematical models as artefacts for research: Felix Klein and the case of Kummer surfaces. *Mathematische Semesterberichte*, *60*, 1–24.
59. Saccheri, G., (2014). *Euclid Vindicated from Every Blemish: Edited and Annotated by Vincenzo De Risi. Translated by GB Halsted and L. Allegri*. Springer, 1, 1-10.
60. Shapiro, S., (2005). Categories, structures, and the Frege-Hilbert controversy: The status of meta-mathematics. *Philosophia Mathematica*, *13*(1), 61–77.
61. Singh, P. K., (2021). Anti-geometry and neutrogeometry characterization of non-Euclidean data. *Journal of Neutrosophic and Fuzzy Systems*, *1*(1), 24–4.
62. Sterrett, S. G., (1994). *Frege and Hilbert on the Foundations of Geometry, 1, 1-25.*
63. Stojanović, S., Pavlović, V., & Janičić, P., (2011). A coherent logic-based geometry theorem prover capable of producing formal and readable proofs. In: *Automated Deduction in Geometry: 8th International Workshop, ADG 2010, Munich, Germany, Revised Selected Papers 8* (pp. 201–220). Springer Berlin Heidelberg.
64. Toussaint, M., (2015). Logic-geometric programming: An optimization-based approach to combined task and motion planning. In: *IJCAI* (pp. 1930–1936).
65. Ungar, A. A., (2008). *Analytic Hyperbolic Geometry and Albert Einstein's Special Theory of Relativity, 1, 1-14.*
66. Wilcox, D. A., & Xie, Y., (2007). Predicting wetland plant community responses to proposed water-level-regulation plans for Lake Ontario:

GIS-based modeling. *Journal of Great Lakes Research, 33*(4), 751–773.

67. Wilke, G., & Frank, A. U., (2010). Tolerance geometry: Euclid's first postulate for points and lines with extension. In: *Proceedings of the 18th SIGSPATIAL International Conference on Advances in Geographic Information Systems* (pp. 162–171).
68. Winer, M. L., & Battista, M. T., (2022). Investigating students' proof reasoning: Analyzing students' oral proof explanations and their written proofs in high school geometry. *International Electronic Journal of Mathematics Education, 17*(2).
69. Withers, J. W., (1905). *Euclid's Parallel Postulate: Its Nature, Validity, and Place in Geometrical Systems.* Open Court Publishing Company, 1, 1-24.
70. Yaman, O., Karaköse, M., Aydın, İ., & Akın, E., (2014). Detection of pantograph geometric model based on fuzzy logic and image processing. In: *2014 22nd Signal Processing and Communications Applications Conference (SIU)* (pp. 686–689). IEEE.

CHAPTER 8

Applications of Geometry in the Real World

Contents

8.1. INTRODUCTION

Applications of geometry extend far and wide into the practical, tangible realms of our daily lives. In the real world, geometry is a powerful tool for understanding, designing, and optimizing various aspects of our environment, from architectural phenomena to intricate technological systems. Through geometry, one can gain insights into the shapes, sizes, and relationships that govern the physical world, enabling us to solve complex problems and create innovations that shape our modern existence.

The applications of geometry in the real world are diverse. From the symmetrical phenomena of architectural design to the intricate calculations powering GPS technology, geometry plays a pivotal role in every field and industry. It is the science of spatial relationships, angles, proportions, and measurements, and its influence is pervasive, linking fields as varied as engineering, art, medicine, and environmental science (Demargne et al., 2014).

This chapter will study the tangible and measurable ways mathematical discipline shapes our daily experiences. From the precise calculations that go into constructing safe and aesthetically pleasing buildings to the geometric principles that support our understanding of the natural world, geometry is the silent force that guides our actions, decisions, and innovations (Purnama et al., 2014).

8.2. ARCHITECTURE AND DESIGN

Geometry provides architects and designers the tools to conceptualize, plan, and create functional and visually captivating structures. From calculating the precise dimensions of a building's foundation to optimizing space distribution within interior layouts, geometry's role is fundamental. It empowers architects to benefit from the features of symmetry, proportion, and angles to craft structures that balance with the natural world while meeting the needs of the inhabitants (Purnama et al., 2014).

8.2.1. Symmetry in Architectural Design

Symmetry is a hallmark of architectural aesthetics. Architects use the power of symmetry to create visually balanced structures. One of the most prevalent forms of symmetry in architecture is bilateral symmetry, where a building is divided into two mirror-image halves. This design approach imparts a sense of balance and equilibrium to structures (Hwang et al., 2015).

Additionally, rotational symmetry plays a role in architectural enhancement. Elements like arches, domes, and vaulted ceilings often exhibit rotational symmetry. For instance, the dome of the Pantheon in Rome showcases perfect rotational symmetry, which serves an aesthetic purpose and contributes to structural stability (Hwang et al., 2015).

- Ø **Example:** The equation for rotational symmetry of a building's facade can be expressed as:

$$R = 360^{\circ} / n$$

where; is the number of identical rotational positions.

8.2.2. Proportions in Building Structures

Proportions are fundamental to architectural design, influencing the dimensions and scale of buildings.

The golden ratio, represented by the mathematical constant phi (Φ), is a key proportion architects frequently employ. It dictates that the ratio of the whole to its larger part is the same as that of the larger to the smaller part. Mathematically, $\Phi = (1 + \sqrt{5})/2$, approximately equal to 1.618.

Architects often use the golden ratio to determine the proportions of building components such as columns, windows, and façades. This ratio is believed to create aesthetically pleasing and harmonious compositions. For example, the Parthenon in Athens, an architectural masterpiece, incorporates the golden ratio in its design, contributing to its visual appeal (Wakamatsu et al., 2005).

- Ø **Example:** To calculate the golden ratio in a building's proportions, you can use the equation:

$$f = \frac{a+b}{a}$$

where; and represent different segments of a building's dimensions.

8.2.3. Geometry in Facade Design

Architects employ geometric patterns, motifs, and tessellations to craft intricate and captivating building exteriors. Geometric tessellations, or arrangements of shapes that completely cover a surface with no overlaps or gaps, are commonly used to create visually engaging facades (Lai & Chien, 2016).

One classic example is the Alhambra in Spain, renowned for its exquisite Islamic geometric patterns. These patterns, constructed with mathematical precision, decorate the palace's walls, ceilings, and archways. The intricate connection of polygons and symmetrical arrangements transforms the Alhambra into a geometric masterpiece.

Ø **Example:** The equation for finding the surface area of a complex facade can be given as:

$$A = \int \int_{D} \sqrt{1 + \left(\frac{\partial z}{\partial x}\right)^2 + \left(\frac{\partial z}{\partial y}\right)^2}\, dx\, dy$$

where; $\left(\frac{\partial z}{\partial x}\right) and \left(\frac{\partial z}{\partial y}\right)$ represent the slope of the facade surface.

8.3. GEOSPATIAL GEOMETRY IN NAVIGATION

Geospatial geometry, a branch of geometry that deals with objects and measurements on the Earth's surface, plays a pivotal role in navigation. From determining one's location using latitude and longitude to projecting efficient routes across the globe, geometry forms the basis of precise and efficient navigation techniques (Retnawati et al., 2017).

8.3.1. Latitude and Longitude

Latitude and longitude are fundamental concepts in geospatial geometry. They provide a global coordinate system that enables accurate location identification (Gecu-Parmaksiz & Delialioglu, 2019).

Tip

When navigating using latitude and longitude coordinates, remember that latitude lines run horizontally, measuring north and south, while longitude lines run vertically, measuring east and west. This understanding is crucial for precise navigation.

Latitude, measured in degrees north or south of the equator, ranges from –90° (South Pole) to +90° (North Pole). Longitude, measured in degrees east or west of the Prime Meridian (located at 0°), spans from –180° to +180°. Together, latitude and longitude uniquely specify any point on the Earth's surface.

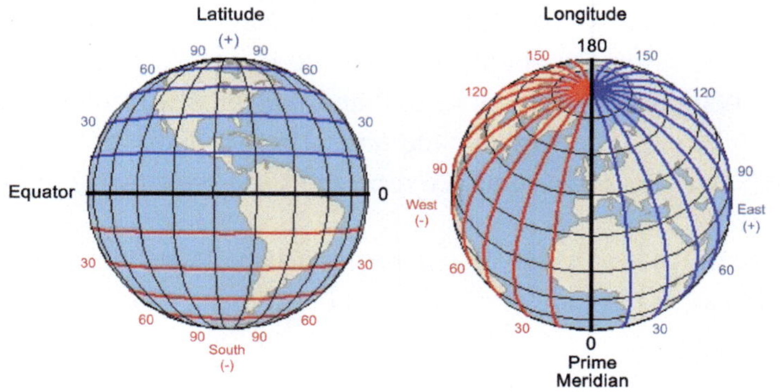

Figure 8.1. Illustration of latitude and longitude in geospatial geometry.

Source: Learnz, Creative Commons License.

Ø **Example:** To convert latitude to radians, use the equation:

$$\text{Latitude in Radians} = \frac{p}{180} \times \text{Latitude in Degrees}$$

8.3.2. Great Circle Routes and Shortest Distances

The shortest distance between two points on the surface of the Earth is not typically a straight line on a flat map. Instead, it follows the path of a great circle, the largest circle that can be drawn on a sphere (Gecu-Parmaksiz & Delialioglu, 2019).

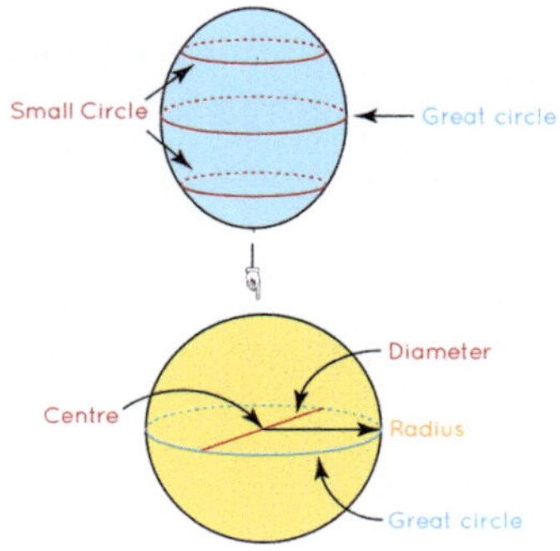

Figure 8.2. Schematic of the great circle in geospatial geometry.

Source: CueMath, Creative Commons License.

Geometrically, great circle routes represent the shortest distance between two locations on the Earth's surface. Navigators and pilots utilize these routes to conserve fuel and time. The concept of a great circle can be illustrated by slicing a sphere with a plane that passes through its center, resulting in a circle. This circle is the largest possible circle on the sphere, and the path along its circumference represents the great circle route.

Ø **Example:** The haversine formula calculates the distance (d) between two points on a sphere using their latitudes and longitudes:

$$d = 2r \cdot \arcsin\left(\sqrt{\sin^2\left(\frac{\text{Ä}q}{2}\right) + \cos(q_1)\cos(q_2)\sin^2\left(\frac{\text{Ä}f}{2}\right)}\right)$$

where; (r) is the radius of the Earth; $(\text{Ä}q)$ is the difference in longitudes; and $(\text{Ä}f)$ is the difference in latitudes.

8.3.3. GPS Technology and Its Geometry

The Global Positioning System (GPS) relies on principles of geometry to determine precise locations on Earth. GPS satellites orbiting the Earth emit signals containing the satellite's position and time. GPS receivers on the ground receive signals from multiple satellites and, through trilateration, calculate their exact position (Bočková et al., 2020).

Trilateration is a geometric method used to determine an object's location based on its distance from three or more known points. In the case of GPS, these known points are the satellites. By measuring the time, it takes for signals from multiple satellites to reach the GPS receiver, the receiver can calculate its precise location using geometric principles.

Ø **Example:** The equation for calculating the distance (d) between a GPS satellite and a receiver is:

$$d = \frac{c \cdot \text{Ä}t}{2}$$

where; (c) is the speed of light and $(\text{Ä}t)$ is the time delay.

8.4. GEOMETRY IN ART AND AESTHETICS

Geometry has had a profound impact on art and aesthetics. It serves as a fundamental framework for understanding and creating art in various forms. The section explores the intricate relationship between geometry and artistic expression, from using geometric shapes and patterns to applying the Golden Ratio, perspective, spatial illusions, and mathematical precision (Yang et al., 2022).

8.4.1. Geometric Shapes and Patterns in Art

Geometric shapes and patterns are prevalent in art across cultures and periods. Artists have long recognized the aesthetic appeal and qualities of geometric forms.

Circles, squares, triangles, and other geometric shapes are often used as building blocks in artistic compositions (Kim et al., 2014).

For example, Dutch artist Piet Mondrian's works are renowned for using simple geometric shapes, particularly rectangles and primary colors, to convey harmony and balance. Geometric patterns, such as tessellations, have been employed in mosaic art and Islamic architecture to create visually stunning designs that repeat geometric motifs seamlessly.

Figure 8.3. Geometric patterns in Islamic art.

Source: Architectural Digest, Creative Commons License.

8.4.2. The Golden Ratio and Its Aesthetic Significance

The Golden Ratio, an irrational number, has captivated artists, architects, and designers for centuries due to its aesthetic significance. In art, the Golden Ratio is associated with attractive proportions and balance (Joye, 2011).

Remember

The Golden Ratio, often represented by the Greek letter Phi (Φ), is approximately equal to 1.6180339887. This irrational number plays a significant role in art, design, and aesthetics, as it's believed to represent ideal proportions that are visually pleasing.

Mathematically, two quantities are in the Golden Ratio if the ratio of their sum to the larger quantity is equal to the ratio of the larger quantity to the smaller one, leading to the equation:

$$\frac{a+b}{a}=\frac{a}{b}=f$$

Artists use the Golden Ratio to determine the dimensions and proportions of their creations. It is often seen in the dimensions of famous artworks, such as Leonardo da Vinci's "Vitruvian Man," where the human body is divided into sections that adhere to the Golden Ratio.

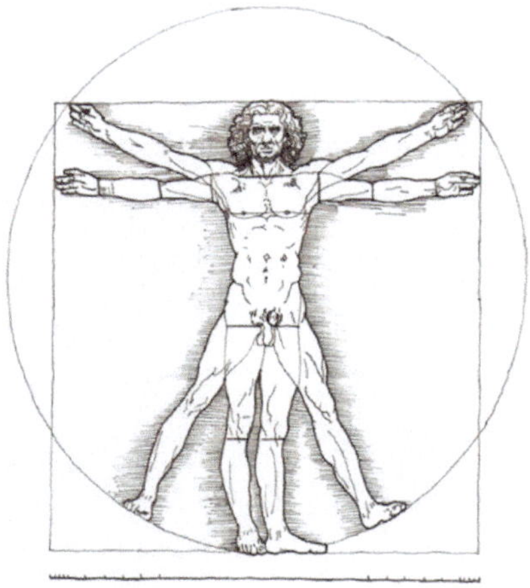

Figure 8.4. Leonardo da Vinci's "Vitruvian Man" is an example of the golden ratio.

Source: Anand Prahlad, Creative Commons License.

Ø **Example:** To express the golden ratio (f) algebraically, you can use the equation:

$$f=\frac{1+\sqrt{5}}{2}$$

8.4.3. Perspective and Spatial Illusions in Painting

Geometry plays a crucial role in portraying three-dimensional space on a two-dimensional canvas. Artists employ perspective techniques to create the illusion of depth and realism in their paintings. Linear perspective, for instance, uses geometric principles to represent objects as they appear in space. One-point perspective involves drawing objects as they recede into the distance along a single vanishing point. Two-point perspective incorporates two vanishing points, providing a sense of depth from different angles. By applying geometric rules, artists can accurately depict how objects converge and diminish in size with distance, creating convincing spatial illusions (Maré, 2018).

- Ø **Example:** The linear perspective equation for drawing receding parallel lines:

$$x' = \frac{x}{d}$$

8.4.4. Mathematical Precision in Sculpture and Architecture

Sculpture and architecture use mathematical precision to achieve structural stability and aesthetic features. Geometric principles guide the construction of sculptures and buildings, ensuring they are visually appealing and structurally sound (Gropius & Kiese, 2016). The Parthenon in Athens, for example, exhibits a high degree of geometric precision in its design, with columns that taper slightly to create an optical illusion of straightness. In sculpture, artists like Constantin Brâncuşi have incorporated geometric elements. Brâncuşi's "Bird in Space" series features sleek, elongated forms that capture the essence of flight and motion through geometric abstraction.

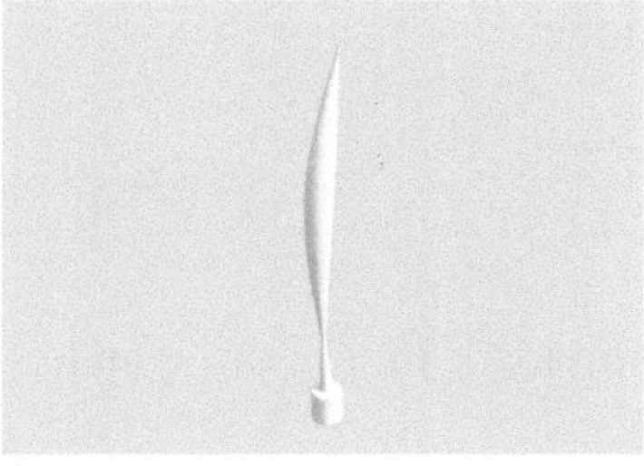

Figure 8.5. Illustration of Brâncuşi's "bird in space."

Source: Sketchfab, Creative Commons License.

In summary, geometry serves as a basis for artistic expression and aesthetics. Artists leverage geometric shapes, patterns, and the Golden Ratio to create visually captivating compositions. Perspective techniques rooted in geometry enable the portrayal of depth in painting, while mathematical precision supports the stability and beauty of sculptures and architectural marvels. Integrating geometry into art and aesthetics enriches our visual experiences and demonstrates the profound interplay between mathematics and creative expression (Pallasmaa et al., 1996).

8.5. ENGINEERING AND STRUCTURAL GEOMETRY

Geometry plays a fundamental role in engineering, particularly in structural engineering, where the design and analysis of various structures depend on geometric principles. This section will discuss the application of geometry in engineering, focusing on how geometric considerations influence the design of trusses, arches, and bridges, as well as how engineers use geometric optimization to create efficient and safe structures.

8.5.1. Geometric Principles in Structural Engineering

Structural engineers rely on geometric principles to ensure the stability, strength, and safety of buildings, bridges, and other infrastructure. Geometry is the backbone of structural design, governing how forces are distributed, materials behave, and loads are transmitted (Rebecchi, 2021). One fundamental concept is the geometric arrangement of structural members, such as beams and columns, to form stable frameworks. Engineers use geometric shapes like triangles and rectangles to design trusses and frames, which are essential for supporting loads and preventing structural failure.

- Ø **Example:** The equation for calculating the tension in a cable of a suspension bridge:

$$T = \frac{W}{2}\sqrt{\frac{L^2}{4} + h^2}$$

where; (W) is the total load; (L) is the span length; and (h) is the sag.

8.5.2. Trusses, Arches, and Load Distribution

Trusses are a common structural element used in engineering, and geometry heavily influences their design. A truss consists of a series of interconnected triangles, which are geometrically stable shapes. Engineers arrange these triangles to efficiently distribute loads and resist deformation (Staples, 2021). For example, in constructing a roof truss, the geometric arrangement of the triangular components helps evenly distribute the weight of the roof and any additional loads, such as snow or wind. By analyzing the forces acting on each member using principles like static equilibrium and the method of joints, engineers can optimize the dimensions and angles of truss components to ensure structural integrity.

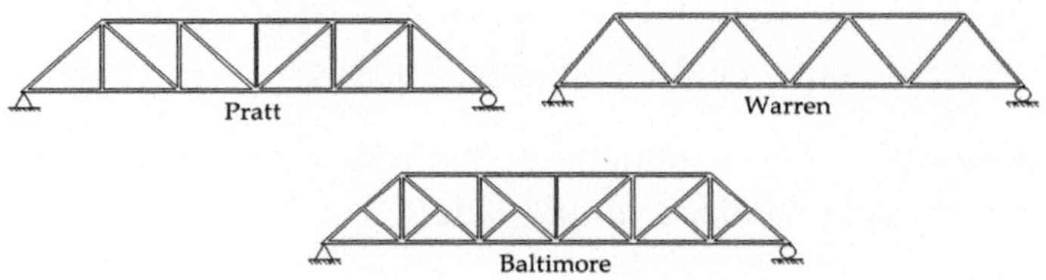

Figure 8.6. Geometry types of trusses.

Source: Arturo Gonzalez, Creative Commons License.

Arches are another geometrical aspect of structural engineering. The semicircular shape of an arch efficiently transfers loads to its supports, making it a stable choice for architectural and engineering applications. The geometry of the arch is crucial in calculating the thrust forces and ensuring that the arch remains stable under the applied loads.

- Ø **Example:** The equation for calculating the maximum load (F) a triangular truss can bear:

$$F = \frac{3}{2} \cdot \frac{P}{\cos(q)}$$

where; (P) is the applied force; and (q) is the angle between the force and the truss.

8.5.3. Geometric Considerations in Bridge Design

Bridges represent a complex connection between geometry and engineering. Engineers must consider various geometric factors, including span length, arch shape, and the distribution of loads, to design safe and efficient bridges (Ferry et al., 2011).

One key geometric concept in bridge design is the span-to-rise ratio, which determines the arch shape. For instance, a bridge with a larger span relative to its rise may adopt a flatter arch shape, while a smaller span-to-rise ratio might result in a steeper arch. The choice of arch shape directly impacts the bridge's aesthetic appearance and structural stability.

Load distribution across a bridge is another critical geometric consideration. Engineers use principles of statics and geometry to calculate how loads, including the weight of vehicles and pedestrians, are distributed among the bridge's supports. It ensures that no part of the bridge experiences excessive stress or deformation (Laborde, 2002).

- Ø **Example:** The equation for the maximum deflection (d) of a beam bridge under a point load:

$$d = \frac{P \cdot L^3}{48 \cdot E \cdot I}$$

8.5.4. Examples of Geometric Optimization in Engineering

Geometric optimization is a powerful tool in engineering, allowing engineers to design efficient and cost-effective structures. This optimization often involves finding the ideal geometric parameters that minimize materials usage while meeting safety and performance requirements. One example is the optimization of beam cross-sectional shapes. Engineers use geometric optimization techniques to determine beams' optimal shape and dimensions, minimizing material usage while ensuring they can withstand the required loads. It reduces material costs and contributes to sustainable and eco-friendly design (Wang & Wen, 2004). Engineers use geometric optimization in aircraft design to shape wings for maximum aerodynamic efficiency. Adjusting parameters such as wingspan, aspect ratio, and airfoil shape can reduce drag, improve fuel efficiency, and enhance overall performance.

- Ø **Example:** The optimization of a wing's lift-to-drag ratio (L/D) in aerospace design:

$$L/D = \frac{\text{Lift}}{\text{Drag}}$$

where; lift and drag are functions of wing geometry and airfoil shape.

In conclusion, the geometric optimization of structural components further enhances the sustainability and functionality of engineered systems.

8.6. GEOMETRY IN MEDICINE AND MEDICAL IMAGING

Geometry plays a crucial role in medicine, from aiding in diagnostics to assisting in surgical procedures. This section explores geometry's applications in medicine and medical imaging, highlighting how geometric models, medical imaging techniques, and surgical procedures are intertwined in the healthcare industry (Saha et al., 2015).

8.6.1. The Role of Geometry in Medical Diagnostics

Geometry is integral to the accurate diagnosis of various medical conditions. For instance, geometric measurements are essential when considering the shape and size of anatomical structures or lesions within the body. Physicians and radiologists rely on geometric parameters to assess the severity of a condition or track changes over time (Goodchild, 2009).

One common application of geometry in diagnostics is tumor size and growth measurement in oncology. Using geometric principles to determine the volume and dimensions of tumors, healthcare professionals can make informed decisions about treatment strategies and monitor a patient's response to therapy.

- Ø **Example:** The calculation of the angle (q) of deflection in medical ultrasound imaging:

$$q = \arcsin\left(\frac{v_t}{v_p}\right)$$

8.6.2. Geometric Models in Anatomy and Physiology

Geometry provides a framework for modeling and understanding complex anatomical and physiological systems. Geometric models help researchers and clinicians visualize and analyze the intricate structures and functions of the human body (Abdelfatah, 2011).

For example, studying blood flow in arteries and veins involves geometric modeling. Scientists use fluid dynamics principles, which rely heavily on geometry, to simulate blood flow patterns and assess the impact of arterial narrowing or blockages. This information aids in the diagnosis and treatment of cardiovascular diseases.

Geometric models also play a role in understanding bone structure and biomechanics. Orthopedic surgeons use these models to plan joint replacements and fracture repairs, ensuring that implants fit precisely and restore normal joint function.

8.6.3. Medical Imaging Techniques: CT, MRI, and Ultrasound

Medical imaging techniques heavily depend on geometry to produce detailed images of internal structures. Three primary imaging methods—computed tomography (CT), magnetic resonance imaging (MRI), and ultrasound—rely on geometric principles to capture and reconstruct images (Toutin, 2011). CT imaging employs a series of X-ray measurements taken from different angles around the patient to create cross-sectional images of the body. These images are then reconstructed using geometric algorithms. The accuracy of CT images relies on precise geometric calculations.

MRI, conversely, uses strong magnetic fields and radio waves to generate images of soft tissues. Geometric principles guide the formation of these images by translating signals from protons in water molecules into spatial information. MRI's ability to capture fine anatomical details has revolutionized medical diagnostics.

Ultrasound imaging relies on the geometry of sound waves. Ultrasound machines create real-time images by emitting high-frequency sound waves into the body and analyzing the returning echoes. Geometric principles are essential for accurately locating structures and assessing dimensions during ultrasound examinations, making it a valuable tool in obstetrics, cardiology, and more.

8.6.4. Surgical Procedures and Geometry

Geometry also plays a crucial role in surgical planning and procedures. Surgeons use geometric principles to navigate intricate anatomical structures, plan incisions, and make precise interventions.

One notable application is minimally invasive surgery, where surgeons make small incisions and use specialized tools to access and treat internal organs. Geometry aids in visualizing the surgical field using three-dimensional models generated from preoperative imaging data. These models guide surgeons in planning the best approach, avoiding critical structures, and achieving optimal outcomes with minimal tissue damage (Toutin, 2011).

In neurosurgery, where precision is paramount, geometric principles help surgeons accurately target specific brain regions. Techniques like stereotactic surgery use geometric coordinates to map the brain's three-dimensional structure, allowing surgeons to perform delicate procedures such as tumor removal or deep brain stimulation.

8.7. GEOMETRY IN COMPUTER GRAPHICS AND GAMING

Geometry is the backbone of modern computer graphics and gaming, enabling the creation of visually stunning virtual worlds, realistic characters, and immersive experiences. This section will discuss the applications of geometry in computer graphics, from 2D and 3D transformations to geometric modeling in video games and the utilization of geometry in virtual reality.

8.7.1. 2D and 3D Transformations in Computer Graphics

Geometry is essential for transforming objects in two and three dimensions within a computer graphics environment. Transformations include translation (shifting objects), rotation (changing their orientation), and scaling (resizing objects) (Toutin, 2011).

In 2D graphics, transformations are achieved using 2×2 matrices. For example, a rotation matrix can rotate a 2D object around the origin, while a translation matrix can move it to a new location. Combining these transformations allows for complex animations and effects.

Ø **Example:** The equation for scaling a 2D object in computer graphics:

$$x' = s_x \cdot x$$

$$y' = s_y \cdot y$$

where; (s_x) and (s_y) are scaling factors.

In 3D graphics, transformations are performed with 3×3 or 4×4 matrices (homogeneous coordinates) to include translations. It enables the manipulation of 3D objects in space. For instance, 3D transformations are used to create the lifelike movement of characters and objects in 3D games, enhancing their realism.

8.7.2. Geometric Modeling in Video Games

Geometric modeling is a fundamental aspect of video game development. It involves creating 3D models of characters, objects, and environments. These models consist of vertices, edges, and faces that define the object's shape and appearance (Toutin, 2011).

Geometric modeling is particularly important for character design. Artists and designers use 3D modeling software to sculpt characters, complete with detailed geometry for facial features, body proportions, and clothing. These models are then animated using skeleton rigs and mathematical algorithms that dictate movement based on the underlying geometry.

In open-world games, geometric modeling extends to creating vast landscapes with intricate details. Terrain generation algorithms use geometric principles to create realistic mountains, valleys, and rivers. Players can explore these environments, adding depth and immersion to the gaming experience.

- Ø **Example:** The use of Bézier curves to define smooth paths for in-game character animations.

Bézier Curve Equation:

$$\boldsymbol{B}(t)=(1-t)^3\boldsymbol{P}_0+3(1-t)^2t\boldsymbol{P}_1+3(1-t)t^2\boldsymbol{P}_2+t^3\boldsymbol{P}_3$$

where; $(\boldsymbol{B}(t))$ is the point on the Bézier curve at parameter (t); and $(\boldsymbol{P}_0),(\boldsymbol{P}_1),(\boldsymbol{P}_2), and (\boldsymbol{P}_3)$ are control points.

8.7.3. Realistic Rendering and Shading Techniques

Geometry plays a significant role in rendering and shading, determining how light interacts with objects in a virtual environment. Realistic lighting and shading effects are crucial for creating visually convincing graphics.

Ray tracing is a geometric rendering technique that simulates the path of light rays as they interact with surfaces. By tracing the paths of individual rays, ray tracing produces lifelike reflections, refractions, and shadows. This technique relies on geometric calculations to determine how light interacts with the geometry of objects in the scene (Maré, 2018).

Shading models like Phong use geometric normals to calculate how light reflects off surfaces. Normals are vectors perpendicular to surfaces at each point, and their orientation affects the shading of objects. Shading models create smooth or rough surface appearances by adjusting the normals, contributing to realism.

Did You Know?

Realistic rendering in computer graphics often involves ray tracing, a technique that simulates the way rays of light interact with objects in a scene. This technology is used extensively in creating lifelike visuals in movies, video games, and architectural visualization.

- Ø **Example:** Applying the Phong reflection model to calculate surface shading in 3D graphics.

Phong Reflection Model Equation:

$$I = k_a I_a + k_d I_d (\boldsymbol{N} \cdot \boldsymbol{L}) + k_s I_s (\boldsymbol{R} \cdot \boldsymbol{V})^n$$

where; (I) is the intensity of light; $(k_a),(k_d)$, and (k_s) is the ambient, diffuse, and specular reflection coefficients $(I_a),(I_d)$, and (I_s) are the intensities of ambient, diffuse, and specular light sources; $(\boldsymbol{N})$ is the surface normal vector; $(\boldsymbol{L})$ is the vector to the light source; $(\boldsymbol{R})$ is the reflection vector; (V) is the view vector; and (n) is the shininess exponent.

8.7.4. Application of Geometry in Virtual Reality

Virtual reality (VR) relies heavily on geometry to create immersive experiences. VR headsets use geometric principles to track the user's head movements and adjust the virtual environment, providing a sense of presence within the digital world (Kim et al., 2014).

Geometric algorithms ensure the virtual environment aligns with the user's perspective, creating a seamless and immersive experience. These algorithms calculate how objects in the virtual world should be rendered based on the user's position and orientation, adjusting the geometry in real-time.

- Ø **Example:** The use of 3D coordinate transformations to position virtual objects in a virtual reality environment.

3D coordinate transformation equations (translation, rotation, scaling):

$$T = \begin{bmatrix} 1 & 0 & 0 & \Delta x \\ 0 & 1 & 0 & \Delta y \\ 0 & 0 & 1 & \Delta z \\ 0 & 0 & 0 & 1 \end{bmatrix}$$

$$(\theta) = \begin{bmatrix} 1 & 0 & 0 & 0 \\ 0 & \cos(\theta) & -\sin(\theta) & 0 \\ 0 & \sin(\theta) & \cos(\theta) & 0 \\ 0 & 0 & 0 & 1 \end{bmatrix}$$

$$R_y(\theta) = \begin{bmatrix} \cos(\theta) & 0 & \sin(\theta) & 0 \\ 0 & 1 & 0 & 0 \\ -\sin(\theta) & 0 & \cos(\theta) & 0 \\ 0 & 0 & 0 & 1 \end{bmatrix}$$

$$R_z(\theta) = \begin{bmatrix} \cos(\theta) & -\sin(\theta) & 0 & 0 \\ \sin(\theta) & \cos(\theta) & 0 & 0 \\ 0 & 0 & 1 & 0 \\ 0 & 0 & 0 & 1 \end{bmatrix}$$

$$S = \begin{bmatrix} S_x & 0 & 0 & 0 \\ 0 & S_y & 0 & 0 \\ 0 & 0 & S_z & 0 \\ 0 & 0 & 0 & 1 \end{bmatrix}$$

where; $(\boldsymbol{T})$ represents translation; $\boldsymbol{R}_x(q), \boldsymbol{R}_y(q)$; and $\boldsymbol{R}_z(q)$ represent rotation about the x, y, and z axes, respectively, and $(\boldsymbol{S})$ represents scaling.

In conclusion, geometry is a foundational element of computer graphics and gaming.

It underpins 2D and 3D transformations, geometric modeling, realistic rendering, shading techniques, and the immersive world of virtual reality. The applications of geometry in these fields result in visually stunning graphics, lifelike animations, and captivating gaming experiences, making it an essential tool for game developers and graphics professionals.

8.8. ENVIRONMENTAL GEOMETRY AND SUSTAINABILITY

Geometry, often associated with abstract mathematical concepts, may seem disconnected from the natural world. However, in environmental conservation and sustainability, geometry plays a significant role. The chapter explores the applications of geometry in environmental preservation, the influence of fractal geometry in understanding natural phenomena, sustainable architecture, urban planning, and geometric approaches to ecological preservation (Yang et al., 2022).

- Ø **Example:** The calculation of the ecological footprint of a sustainable urban development project using geometric analysis.

Ecological Footprint Equation:

$$EF = \frac{\text{Total Resource Consumption}}{\text{Biocapacity of the Ecosystem}}$$

8.8.1. Geometry in Environmental Conservation

Environmental conservation requires a deep understanding of spatial relationships and geographical data. Geometry provides the tools to analyze landscapes, map ecosystems, and plan conservation efforts effectively. Geographic information systems (GIS), heavily reliant on geometric principles, allow researchers and policymakers to collect, analyze, and visualize data related to the environment.

One key application is spatial analysis, which uses geometric operations to study the proximity of habitats, identify critical areas for conservation, and assess the impact of human activities on ecosystems. For instance, buffer zones around protected areas are defined using geometric calculations to reduce the negative effects of human encroachment (Yang et al., 2022).

8.8.2. Fractal Geometry in Natural Phenomena

Fractal geometry, a branch of mathematics that deals with self-replicating geometric patterns, has found application in understanding various natural phenomena.

Many natural objects like coastlines, trees, and clouds exhibit fractal-like characteristics. These patterns can be described mathematically using fractal geometry (Bi et al., 2015).

The concept of fractal dimension is particularly valuable in quantifying the complexity of natural forms. It allows scientists to measure how irregular and intricate natural objects are. For instance, a coastline may have a non-integer fractal dimension, indicating its convoluted and self-replicating nature. This dimension helps in modeling and predicting changes in natural systems.

Ø **Example:** Applying the Mandelbrot set equation to generate realistic terrain landscapes.

Mandelbrot Set Equation:

$$Z_{n+1} = Z_n^2 + C$$

where; (Z_n) is a complex number; and (C) is a constant. The visualization of the Mandelbrot set results in intricate fractal patterns resembling natural landscapes.

8.8.3. Sustainable Architecture and Urban Planning

Geometry plays a pivotal role in sustainable architecture and urban planning. Sustainable design seeks to minimize the negative impact of buildings and cities on the environment while maximizing efficiency and comfort. Geometric principles contribute to achieving these goals.

Architects use geometry to optimize building designs for energy efficiency. Passive solar design, for example, relies on geometric calculations to determine the orientation and placement of windows, allowing buildings to harness natural sunlight for heating and lighting. Moreover, geometric analysis of building materials helps architects select sustainable options that reduce waste and energy consumption (Rahman, 2020).

Urban planners employ geometric concepts to create environmentally friendly cities. Efficient public transportation systems are designed using network theory, a branch of geometry, to minimize travel distances and reduce carbon emissions. The layout of green spaces, bike lanes, and pedestrian walkways is geometrically planned to promote sustainable mobility and reduce car dependence.

8.8.4. Geometric Approaches to Ecological Preservation

Ecological preservation aims to protect and restore natural ecosystems. Geometry plays a role in ecological studies by aiding in the measurement of biodiversity, the design of wildlife corridors, and the restoration of degraded landscapes.

Species distribution modeling, a geometric approach, uses mathematical equations and spatial data to predict the potential habitats of various species. This information is crucial for conservation efforts as it guides the establishment of protected areas and the creation of wildlife corridors, geometrically designed pathways that connect fragmented habitats, allowing species to migrate and maintain genetic diversity (Retnawati et al., 2017).

Additionally, geometric principles are applied in landscape restoration projects. Techniques like contour farming and terracing use geometric shapes to control soil erosion and water runoff. These methods prevent land degradation, promote sustainable agriculture, and protect water resources.

8.9. GEOMETRY IN EVERYDAY LIFE

Geometry is not confined to the pages of textbooks or the walls of classrooms. It permeates every aspect of our daily lives, often in ways we may not consciously recognize. In this section, we explore how geometry influences our homes, gardens, the products we use, and even the sports and recreational activities we enjoy (Lai & Chien, 2016).

8.9.1. Geometry in Home Design and Interior Layouts

When designing a home or planning the layout of interior spaces, architects, and homeowners alike rely on geometric principles to create functional and aesthetically pleasing environments (Wakamatsu et al., 2005).

1. **Room Dimensions:** Geometry influences room dimensions and proportions. The Golden Ratio, a mathematical ratio often found in nature and art, creates pleasing room proportions. It helps determine the width, height, and length of spaces in a way that is visually pleasing to occupants.
2. **Furniture Arrangement:** When arranging furniture, homeowners consider geometric factors. Furniture should fit within a room's dimensions while allowing for comfortable movement. The geometry of furniture placement ensures that there is a balance between open spaces and functional areas.
3. **Interior Decorating:** Patterns and designs featuring geometric shapes are prevalent in interior decorating. From wallpaper with tessellating patterns to decorative tiles with intricate geometrical

motifs, geometry adds a sense of order and style to our living spaces.

Figure 8.7. Application of geometry in home design.

Source: Trendir, Creative Commons License.

8.9.2. Practical Geometry in Landscaping and Gardening

Geometry is not limited to the interior of our homes; it extends into our outdoor spaces as well (Cumino et al., 2017).

1. **Garden Design:** Landscapers and gardeners use geometric principles to design and plan outdoor spaces. Rectangular and circular garden beds, pathways, and the arrangement of plants often follow geometric patterns to create visually appealing landscapes.
2. **Irrigation Planning:** The layout of irrigation systems for gardens and lawns relies on geometric calculations to ensure even water distribution. Irrigation pipes and sprinklers are strategically placed based on the geometry of the area to prevent overwatering or dry patches.
3. **Hardscaping:** Elements like patios, walkways, and retaining walls incorporate geometry to ensure stability and durability. The angles and dimensions of these structures are calculated to withstand environmental stresses.

8.9.3. Geometric Considerations in Product Design

Our daily products are designed with geometry to enhance functionality, safety, and aesthetics (Liná Wang, 2014).

1. **Electronic Devices:** The design of electronic devices, such as smartphones and tablets, relies heavily on geometric principles. Engineers use geometry to determine screen dimensions, button placement, and the overall form factor to optimize user experience and aesthetics.
2. **Automotive Design:** Geometry plays a significant role in the design of automobiles. Vehicle safety, aerodynamics, and fuel efficiency are all influenced by geometric considerations. For example, the shape of a car's body is designed using mathematical modeling to reduce drag and improve fuel economy.
3. **Household Items:** From kitchen appliances to furniture, the shape and dimensions of household items are carefully considered. Ergonomics, which involves the geometry of human interaction with objects, ensures that products are comfortable and efficient.

8.9.4. Geometry's Role in Sports and Recreation

Geometry profoundly impacts various sports and recreational activities, from the design of sports equipment to the rules of the games themselves.

1. **Sports Equipment:** The design of sports equipment, such as golf clubs, tennis rackets, and soccer balls, relies on geometry. Engineers use mathematical modeling to optimize these items' aerodynamics, weight distribution, and impact characteristics.
2. **Game Strategy:** Many sports involve strategic positioning and angles. In billiards, players use geometry to calculate the rebound angles of balls off the cushions. Similarly, basketball players consider the angles and trajectories of shots to increase the likelihood of scoring.
3. **Outdoor Activities:** Activities like hiking and camping often require an understanding of geometry. Navigational tools like compasses and maps are based on geometric principles, enabling outdoor enthusiasts to determine direction, distance, and elevation.

In conclusion, geometry is an integral part of our daily lives, influencing everything from the design of our homes to the products we use and the recreational activities we enjoy. Whether we consciously recognize it or not, geometry is pivotal in enhancing functionality, aesthetics, and efficiency in our everyday existence. Understanding and appreciating the geometrical aspects of our world can lead to a deeper appreciation of the precision and elegance that geometry brings to our lives.

Activity 8.1

Imagine you're a product designer tasked with creating a new ergonomic chair. How would you use geometric principles to ensure that the chair is not only comfortable but also aesthetically pleasing? Consider factors like the angles of the seat and backrest, the distribution of pressure points, and the chair's overall shape.

8.10. SUMMARY

This chapter discussed the practical applications of geometry across various fields. It begins with a look at architecture and design, highlighting the importance of symmetry and proportions in building structures. Geospatial geometry in navigation, using latitude and longitude, is explored, emphasizing its role in precise location determination, including GPS technology. The chapter discusses the presence of geometry in art and aesthetics, covering concepts like the Golden Ratio and its significance in art and architecture. Engineering and structural geometry are examined in terms of principles, load distribution, and optimization in design. Additionally, the chapter explores the role of geometry in medicine, computer graphics, environmental conservation, and everyday life, showcasing its wide-ranging impact on our world.

REFERENCES

1. Abdelfatah, H., (2011). A story-based dynamic geometry approach to improve attitudes toward geometry and geometric proof. *ZDM, 43*, 441–450.
2. Bautista, H. M., Reyes, H. C., Esparza, M. A. O., Arteaga, J. M., & Vera-Vasquez, C. G., (2022). *Model for the Development of Virtual Reality Applications for Learning Geometry in Basic Education, 1, 1-15.*
3. Bi, Y., Lv, M., Song, C., Xu, W., Guan, N., & Yi, W., (2015). AutoDietary: A wearable acoustic sensor system for food intake recognition in daily life. *IEEE Sensors Journal, 16*(3), 806–816.
4. Bočková, V., Pavlovičová, G., & Čeretková, S., (2020). Increasing pupils' interest in geometry through mathematical trails. In: *ICERI2020 Proceedings* (pp. 2038–2047). IATED.
5. Borghese, A., Rea, N., Turolla, R., Rigoselli, M., Alford, J. A. J., Gotthelf, E. V., & Sathyaprakash, R., (2021). The X-ray evolution and geometry of the 2018 outburst of XTE J1810-197. *Monthly Notices of the Royal Astronomical Society, 504*(4), 5244–5257.
6. Boto, E., Holmes, N., Leggett, J., Roberts, G., Shah, V., Meyer, S. S., & Brookes, M. J., (2018). Moving magnetoencephalography towards real-world applications with a wearable system. *Nature, 555*(7698), 657–661.
7. Buitrón-Luna, J., & Cabanillas-Carbonell, M., (2021). Influence of augmented reality for the learning of geometric solids in primary school students. In: *2021 IEEE Engineering International Research Conference (EIRCON)* (pp. 1–4). IEEE.
8. Cumino, C., Spreafico, M. L., & Zich, U., (2017). From museum education and entertainment to daily life edutainment: Tools for understanding geometric shapes. In: *ICERI2017 Proceedings* (pp. 4447–4454). IATED.
9. Demargne, A. A., Evans, R., Tiller, P., & Dawes, W. N., (2014). Practical and reliable mesh generation for complex, real-world geometries. In: *52nd Aerospace Sciences Meeting* (p. 0119).
10. Ferry, B., Duclos, M., Burt, L., Therre, P., Le Gall, F., Jaffré, C., & Courteix, D., (2011). Bone geometry and strength adaptations to physical constraints inherent in different sports: Comparison between elite female soccer players and swimmers. *Journal of Bone and Mineral Metabolism, 29*, 342–351.

11. Gecü, Z., & Özdener, N., (2010). The effects of using geometry software supported by digital daily life photographs on geometry learning. *Procedia-Social and Behavioral Sciences, 2*(2), 2824–2828.
12. Gecu-Parmaksiz, Z., & Delialioglu, O., (2019). Augmented reality-based virtual manipulatives versus physical manipulatives for teaching geometric shapes to preschool children. *British Journal of Educational Technology, 50*(6), 3376–3390.
13. Goodchild, M. F., (2009). Geographic information systems and science: today and tomorrow. *Annals of GIS, 15*(1), 3–9.
14. Gropius, W., & Kiese, D., (2016). The Bauhaus and how do we build decent, beautiful, and inexpensive housing? *West 86th: A Journal of Decorative Arts, Design History, and Material Culture, 23*(1), 102–124.
15. Gupta, R., & Xie, H., (2018). Nanoparticles in daily life: Applications, toxicity, and regulations. *Journal of Environmental Pathology, Toxicology, and Oncology*, *37*(3), 209-230.
16. Hwang, W. Y., Lin, L. K., Ochirbat, A., Shih, T. K., & Kumara, W. G. C. W., (2015). Ubiquitous geometry: Measuring authentic surroundings to support geometry learning of the sixth-grade students. *Journal of Educational Computing Research, 52*(1), 26–49.
17. Hwang, W. Y., Purba, S. W. D., Liu, Y. F., Zhang, Y. Y., & Chen, N. S., (2018). An investigation of the effects of measuring authentic contexts on geometry learning achievement. *IEEE Transactions on Learning Technologies, 12*(3), 291–302.
18. Ito, S., & Dechant, A., (2020). Stochastic time evolution, information geometry, and the Cramér-Rao bound. *Physical Review X, 10*(2), 021056.
19. Joye, Y., (2011). A review of the presence and use of fractal geometry in architectural design. *Environment and Planning B: Planning and Design, 38*(5), 814–828.
20. Khan, M., Gulistan, M., Yaqoob, N., Khan, M., & Smarandache, F., (2019). Neutrosophic cubic einstein geometric aggregation operators with application to multi-criteria decision-making method. *Symmetry, 11*(2), 247.
21. Kim, S. C., Han, B. K., Seo, J., & Kwon, D. S., (2014). Haptic interaction with virtual geometry on robotic touch surface. In: *SIGGRAPH Asia 2014 Emerging Technologies* (pp. 1–3).

22. Laborde, C., (2002). Integration of technology in the design of geometry tasks with Cabri-geometry. *International Journal of Computers for Mathematical Learning, 6*, 283–317.
23. Lai, C. Y., & Chien, C. C., (2016). Geometry-induced memory effects in isolated quantum systems: Cold-atom applications. *Physical Review Applied*, *5*(3), 034001.
24. LináWang, Z., (2014). Triboelectric nanogenerators as new energy technology and self-powered sensors – Principles, problems, and perspectives. *Faraday Discussions, 176*, 447–458.
25. Maré, E. A., (2018). A rhetorical analysis of the facades of buildings designed by Leon Battista Alberti. *South African Journal of Art History, 33*(2), 88–114.
26. Martin, J., Cantelli, A., Paola, C., Blum, M., & Wolinsky, M., (2011). Quantitative modeling of the evolution and geometry of incised valleys. *Journal of Sedimentary Research, 81*(1), 64–79.
27. Pallasmaa, J., Gozak, A., Mel'nikov, K. S., & Cooke, C., (1996). *The Melnikov House, Moscow (1927–1929): Konstantin Melnikov, 1, 1-14.*
28. Purnama, J., Andrew, D., & Galinium, M., (2014). Geometry learning tool for elementary school using augmented reality. In: *2014 International Conference on Industrial Automation, Information, and Communications Technology* (pp. 145–148). IEEE.
29. Rahman, K., (2020). Approaches to some induced Einstein geometric aggregation operators based on interval-valued Pythagorean fuzzy numbers and their application. *New Mathematics and Natural Computation, 16*(02), 211–230.
30. Rebecchi, M., (2021). Abstract moving images and expanded architectures. The 1958 world's fair in brussels and the Philips Pavilion. *Exposing the Moving Image: The Cinematic Medium Across World Fairs, Art Museums, and Cultural Exhibitions, 1, 1-18.*
31. Retnawati, H., Arlinwibowo, J., & Sulistyaningsih, E., (2017). The students' difficulties in completing geometry items of national examination. *International Journal, 8*(4), 03.
32. Rohan, T. M., (2007). Challenging the curtain wall: Paul Rudolph's blue cross and blue shield building. *Journal of the Society of Architectural Historians, 66*(1), 84–109.

33. Saha, P. K., Strand, R., & Borgefors, G., (2015). Digital topology and geometry in medical imaging: A survey. *IEEE Transactions on Medical Imaging, 34*(9), 1940–1964.
34. Seff, A., Ovadia, Y., Zhou, W., & Adams, R. P., (2020). Sketchgraphs: A large-scale dataset for modeling relational geometry in computer-aided design. *arXiv Preprint arXiv:2007.08506.*
35. Singh, R. P., (2015). Heritage value of religious sites and built archetypes: the scenario of Hinduism and illustrating the riverfront Varanasi. *Aatmbodh, the Journal of Knowledge of Self, 12*(1), 21–40.
36. Staples, D., (2021). Sydney Opera House, 1973 Sydney, Australia: Architect, Jørn Utzon. In: *Modern Theatres 1950–2020* (pp. 237–248). Routledge.
37. Toutin, T., (2011). State-of-the-art of geometric correction of remote sensing data: A data fusion perspective. *International Journal of Image and Data Fusion*, *2*(1), 3–35.
38. van Thoor, M. T. A., (2020). An extraordinary photograph: Gerrit Rietveld, Mart Stam and El Lissitzky at the Schröder House, 1926. In: *Dutch connections: Essays on International Relationships in Architectural History in Honor of Herman van Bergeijk* (pp. 381–390). Delft University of Technology.
39. Wakamatsu, H., Takahashi, K., & Hirai, S., (2005). Dynamic modeling of linear object deformation based on differential geometry coordinates. In: *Proceedings of the 2005 IEEE International Conference on Robotics and Automation* (pp. 1028–1033). IEEE.
40. Wang, Y., & Wen, L., (2004). Mapping the geometry and geographic distribution of a very low velocity province at the base of the Earth's mantle. *Journal of Geophysical Research: Solid Earth, 109*(B10).
41. Yang, B., Bao, C., Zeng, J., Bao, H., Zhang, Y., Cui, Z., & Zhang, G., (2022). Neumesh: Learning disentangled neural mesh-based implicit field for geometry and texture editing. In: *European Conference on Computer Vision* (pp. 597–614). Cham: Springer Nature Switzerland.

INDEX

S

T